JUST WEEDS

WEEDS

HISTORY, MYTHS, AND USES

PAMELA JONES

WITH ILLUSTRATIONS BY BOB JOHNSON

PRENTICE
HALL
PRESS

NEW YORK·LONDON·TORONTO·SYDNEY·TOKYO·SINGAPORE

PRENTICE HALL PRESS
15 Columbus Circle
New York, New York 10023

PRENTICE HALL PRESS and colophons are registered
trademarks of Simon & Schuster, Inc.

Library of Congress Cataloging-in-Publication Data

Jones, Pamela.
    Just weeds :   history, myths, and uses / Pamela Jones.
        p.   cm.
    Includes bibliographical references and indexes.
    ISBN 0-13-514118-4
    1. Weeds—History.   2. Weeds—Folklore.   3. Weeds—Utilization.
I. Title.
SB611.J65   1991
632'.58—dc20   90-7778
                                                       CIP

Designed by Barbara Cohen Aronica

Manufactured in the United States of America

10   9   8   7   6   5   4   3   2   1

First Edition

With special love
for
my son,
Christopher

CONTENTS

All thirty weeds discussed in this book are listed here in alphabetical order according to their botanical names, together with their most frequently used common names. To help readers find the chapter about a weed whose botanical name they may not be familiar with, the Index of Botanical and Common Names (see page 287) provides a complete cross-reference for the many common names by which each plant is (or has been) known.

INTRODUCTION

When I was a child, a family friend gave me a small book called *My Little Garden*. Delicately colored illustrations showed twelve familiar flowers dressed like pretty children, and accompanying each illustration was an allegorical verse. I remember being regularly moved to tears by two of the representations. In the first, the neatly dressed viola speaks over the garden fence to the lonely dandelion outside, marveling at the dandelion's ability to thrive in such poor soil, untended by a gardener's care. The dandelion, dressed only in shabby sackcloth, is nevertheless radiant under its crown of spring gold and modestly places its life in the hands of God. The second illustration, the last in the book, is named "The Weed." Here again is the dandelion, significantly inside the fence this time. Only now its crown is gone, its hair is lank, its feet are bare, and its sackcloth dress is little more than a rag. Forlorn and ignored, with a single tear of despair on its cheek, the dandelion wrings its hands outside a bed of healthy, colorful flowers, pleading for just a little love.

Of course, it is difficult for me to say whether it is that little book that has always kept me from despising weeds, or whether it was the walks with my father before he died when I was seven, which infected me with his contagious passion for life and all things living. Or it may have been the countless carefree hours my sister and I spent in Mother's tow, foraging in fields and woods, along streams and on hillsides, for many of the wonderful foods on which we grew up.

Over the years since those early times, I've never ceased to be fascinated by the inexhaustible inventiveness of Nature and all its generous offerings. Among the most abundant of these are what some call weeds and others call wild plants or herbs. It's a curious fact that wild plants or herbs are gladly admired by the human race—even protected, if need be, against possible extinction. But let them be called weeds, however erroneously, and even kind, gentle souls turn to violence. Occasionally, I venture to ask an avid weed hunter, "Do you know that's dinner you're throwing away?" But it's no good. The pitying look I receive tells me so.

Weeds are bad; everybody says so, therefore, it must be true. And so we follow a fairly accepted pattern of usage. If we personally buy a plant at a nursery and set it in a garden bed, we recognize it as a garden plant; if it's in a field or along a roadside, we call it a wild flower or a wild plant; if we shake it, dried, out of a jar and into our stewpot, we call it an 'erb or herb; and if it messes up the perfection of our lawn or the vegetable garden, then it's a weed. And weeds must be pursued, dismembered, trampled, and destroyed without mercy.

Even the relatively few books whose titles are bold enough to say the word *weed* often become less lionhearted between their covers: Some admit a shamefaced admiration for the subject; others good-humoredly dismiss as mere folklore the usefulness of certain weeds; and still others guardedly slip the word *herb* into the text. Most books about the subject resolutely favor the strict control, or extermination, of weeds in the interests of farmers and agriculture. Only two books (Edwin Rollin Spencer's *All About Weeds* and Sara B. Stein's *My Weeds*) treat the subject with a fine blend of informed objectivity, good-humored understanding, and affection.

Far more books, however, avoid mentioning *weeds* altogether, and concentrate instead on *wild plants* or *herbs*. Such an approach may be encouraging to all falsely accused weeds, but does little to help the individual in search of real information. General gardening books, if they do not altogether ignore the subject, often include only small sections on weeds, presenting them not unlike the FBI's "most wanted" lists, as the thirty or fifty "most common weeds." Unfortunately, no two such lists agree wholly on the same thirty or fifty weeds, and even a cursory scan quickly reveals that easily two-thirds of all these "most common" weeds cannot be properly condemned to weedhood at all.

In the face of so much ambiguity and ambivalence, and in the absence of a widespread popular concern for the subject, it is not surprising that the killing of "weeds" is big business. It may be said without exaggeration that weeds are among the most costly problems for farmers and gardeners, either in direct expenditures or in labor costs. In fact, almost from the moment the groundhog sees, or does not see, his shadow on February 2 every year, the subject of weeds begins to take up airtime and print space. The subject is not discussed on the basis of this (named) plant or that (named) plant; it is simply "weeds"—collectively. Health reports warn of allergies; forecasts predict both the weather and anticipated weed growth and pollen count; articles advise how to prevent, minimize, survive, and otherwise foil the evil intent of the mean-spirited weeds all around us. And, almost before the last snow has melted, the sales of preemergent and postemergent weed killers go into high gear.

It seems to me, however, that if a subject is worthy of so much public awareness, discussion, economic investment, fear, and contempt, then surely it is also worthy of being suitably identified. Exactly what *is* a weed? Pressed for an answer, most people will approximate the dictionary definition of the word *weed* as meaning something useless, unwanted, invasive, and without economic value. That, alas, leaves unanswered the matter of recognition. Identifying a weed is not like identifying a house, a fishing rod, or a sandwich, the words for all three of which evoke instant mental images.

Ask the same people to name at least three weeds, and the momentary silence becomes deafening. True, such a casual survey will eventually elicit a varied list of most-hated weeds. Not surprisingly, the dandelion will be far in the lead, because, of all weeds, it is probably the most universally recognized. Others may include, but certainly not be limited to, quack grass, stinging nettles, burdock, crabgrass, groundsel, chicory, ragweed, pokeweed, or purslane. All of these are widespread and common in many regions, although they represent only a handful among the hundreds of weeds native to, or naturalized in, North America.

Interestingly, of the ten "weeds" I have named, only two are determinedly useless—at least for now. They are crabgrass and ragweed. Nevertheless, I'm reluctant to condemn even these two to true "weedhood," just in case they represent the as yet undiscovered key to eternal youth, or to a world free from hunger and disease. The remaining eight "weeds," together with countless others, do not meet a weed's primary criterion of uselessness and should, therefore, not be called weeds.

So if they should not be called weeds, why are they? What *should* they be called? And if they're not useless, what purpose *do* they serve? The first question is fairly easy to answer: To the person who planted a lawn with grass seeds and a flower bed with petunia seeds, for instance, the emergence of dandelions, plantain, or burdock comes as an unpleasant surprise. In such circumstances, any one of these plants, especially in large numbers, would be quickly condemned to the dictionary definition of a weed.

The answer to the second question will be yours to give. Should they be "wild plants," "wildlings," or "herbs"? Or should they remain "weeds"? To help you reach a fair verdict, I offer you the evidence in the following pages. Without seeking to influence your decision, I ask only that you satisfy your conscience and your reason.

The answer to the third question is the focus of this book.

Only seldom do most of us take the trouble to give weeds more than a cursory glance or thought. Yet, when we do look at them, we tend to be rather surprised to discover that many of them are exceptionally beautiful. In such expansive moments, we notice the delicate intricacy of Queen Anne's lace, the luminosity of chicory, and the regality of mullein wrapped in its silver-green fur. We might remember how, as children, we could not resist blowing on the fluffy seed heads of dandelions, or plucking the petals off oxeye daisies to find out if our true love was true. If we heard that one of our common weeds was already ancient when dinosaurs roamed the land, would we believe it? Or that the seeds of another of our weeds are carnivorous? Would we imagine that there is written proof that one of the commonest weeds existed 4,000 years ago, or think of clothing ourselves with a weed few people dream of even touching?

It is interesting to note that English is the only major European language that has severed all sense of connectedness between the word and concept of *weed* and other plants. By contrast, the words for weed in French and Spanish mean "bad herb," and in German it is simply "un-herb," while the Italians, with their typical generosity of nature, call it *erbaccia*, the next best thing to *erba*, for herb.

Most of our weeds are the descendants of weeds that migrated to our shores from around the globe—predominantly from Europe and Asia. With many of them came their traditional usage, steeped in folklore, history, myth, and witchcraft. Many of them have been cultivated for centuries, and several have inspired heroic deeds, poetry, or great inventions. They have served as food or medicine, as domestic aids or cosmetics, or as all of these; a few of them are—and have been for centuries—major agricultural crops.

With the age of technology and the developments of the industrial revolution and modern medicine, however, many of the traditions concerning weeds lost much of their influence. Moreover, in North America, the divers tribes of American Indians, the best source of knowledge about the uses of indigenous plants, were forced to leave their homelands and dispersed to reservations. And so, perhaps particularly in America, the lore of wild plants fell out of the public eye, although it was kept alive mainly by the steadfastness of naturalists and traditionalists. In today's growing popular concern with and awareness of our environment and the need to protect it in order to protect ourselves, interest in and the search for herbal knowledge are rapidly increasing. And, as herbalists have known since ancient times, the wild plants we call weeds are an integral part of that knowledge.

In Europe, perhaps because of its age-old herbal traditions, wild plants have never ceased to be incorporated as a normal part of life. Wherever it is possible or advisable to do so, certain edible wild plants are eagerly sought there in spring, both as desirable delicacies and for the richness of their vitamin content. In large parts of Europe, the herbal medicine of homeopaths and herbalists is often practiced side by side with formal medicine, in the belief among many practitioners of both professions that this practice is the only way to heal the *entire* person, physically, emotionally, and spiritually. And in many parts of the world, wild plants continue to play a role in certain religious and other rites, as they did of yore.

In describing many of the uses to which the weeds of my choice have been put, I have relied on the published works of historic, as well as modern, herbalists, botanists, physicians, and foragers. If many of these are European, it is perhaps because the usage of plants for various purposes has been common and continuous in a large part of European life for many centuries and in numerous cultures.

I grew up with some of these traditions, on both sides of the English channel. Consequently, wherever applicable, I have also included my own experiences and those of family members and close friends with some of the weeds. In the Vital Statistics section, I indicate which parts of a given plant are generally used; in a few instances this may be the "whole plant," that is, including the roots; or it may be the "whole herb," in which case the roots are excluded. For me, there is a unique satisfaction whenever I reach for an appropriate weed, if I sense a headache coming on, or need a hand lotion, or want a facial, or feel thirsty. Or, if I crave a particular flavor—like lamb's-quarters, or chickweed, or stinging nettles, or pink "lemonade."

My chief purpose in writing *Just Weeds* is to cast light on the often astounding histories of some of our most common, most despised wild plants, and to inform and entertain with accounts of their myths, folklore, and traditional usages. What *Just Weeds* is *not* intended for is as medical recommendations or advice. On the contrary, I cannot stress enough that greatest care must be taken in approaching the use of *any* unknown plant, no matter how "safe" it is said to be. Moreover, a few of the weeds described in these pages are poisonous or unsafe—and are clearly indicated as such.

Numerous historical remedies have not met with scientific agreement or approval, under scrutiny. This includes those made from plants whose specific name is *officinalis* (or *officinale*), both meaning "of the [apothecary's] workshop," in reference to earlier times, when *officinalis* plants were highly valued and prescribed by the predecessors of modern-day pharmacists. However, modern herbalists and homeopaths, particularly in Europe, continue to employ many traditional treatments.

In contemplating the internal use of any wild plant, caution is most assuredly recommended. Even so, to strike a balance, it is well worth remembering that even milk, than which few foods are deemed safer, can cause allergic reactions; that green potatoes and the leaves of garden rhubarb can be poisonous; and that something as seemingly harmless as apple seeds can cause death.

Therefore, before using any untried plant, the following steps are both wise and essential:

1. Be certain of a plant's accurate *botanical* name—several very different plants may share the same *common* name, and often do.

2. Never gather plants from areas known to have been sprayed with chemical insecticides or herbicides.

3. Eat edible weeds sparingly at first to let the body's system adjust to the change in diet.

4. Never enroll in the school of thought that believes if "a little is good, a lot is better."

5. Always consult a medical or health professional should the need for one be indicated. Pregnant women should *always* consult a physician before taking any herbal preparation.

Lastly—and ideally—I would like to see the word *weed* abolished altogether for being one of the most intolerant, negative words in the English language. To help make this possible, I hope to lure you in these pages on a journey of discovery among some of nature's most extraordinary survivors and heroes. All of them have served humankind in one manner or another, directly or indirectly, in most cases for thousands of years. Of course, I'm biased! In offering my own list of thirty "most common" weeds, I include not only those that are among the most prevalent and most easily recognized but whose "life stories" are among the most intriguing. My objective is simple: to turn your resentment and contempt of them into a curiosity that will lead eventually to their acceptance as a normal part of life, like air and water and sun and earth.

Achillea millefolium

Achillea millefolium

(Y A R R O W)

VITAL STATISTICS

Perennial

COMMON NAMES	Yarrow, Milfoil, Sanguinary, Bloodwort, Stanchgrass, Thousandleaf, Nosebleed, Old Man's Pepper, Soldier's Woundwort, Knight's Milfoil, *Herba Militaris*, Yarroway, Thousand Seal, Carpenter's Weed, Staunchweed, Devil's Nettle, Devil's Plaything, Angel Flower, Bad Man's Plaything, Dog Daisy, Goose Tongue, Melancholy, Sweet Nuts, Wild Pepper, Knight's Balm, Old Man's Mustard, Bunch o' Daisies
USES	Culinary, medicinal, cosmetic
PARTS USED	Whole herb
HEIGHT	To 3 feet, rough, angular, dull grayish green stems
FLOWER	Terminal white, flat-topped, clustered flower heads, composed of miniature daisylike florets; June to October; white to crimson
LEAVES	Feathery, fernlike, dark green, stalkless, clasping stem at base
ROOT TYPE	Woody, white, branched
HABITAT	Fields, roadsides, waste places, sunny stream banks
PROPAGATION	Seeds, cuttings
CONTROL	Frequent cultivation, weeding

Yarrow has been thriving in Eurasia (and in other temperate regions around the world) for at least 60,000 years, according to the carbon dating of fossilized pollen found in Neanderthal burial caves in Iraq. The plant was probably known by ancient Egyptian and Indian practitioners of medicine some 5,000 years ago, as well as by the earliest developers of the Chinese system of medicine not much later. Certainly, the use of dried, stripped yarrow stalks to cast the *I Ching* is an ancient means of divining the future.

However, it was Achilles, the hero of Homer's *Iliad*, who set yarrow in the laps of legend and history during the Trojan War, about 1200 B.C. On the battlefield outside King Priam's city, this handsomest and bravest of celebrated Greek warriors taught his soldiers the virtues of the herb, as he had learned them from the wise Chiron, whose knowledge of the healing art was renowned. Achilles stanched his comrades' bleeding by packing yarrow leaves on their wounds, including those of his good friend Telephus, king of Mysia and son of Hercules.

History or legend, this is how yarrow got the name *Achillea*. In all, there are some eighty species of plants belonging to this genus, all of them yarrows, all of them perennials, and most of them wild plants. The name yarrow itself derives from *gearwe*, the Anglo-Saxon name of the plant; the Dutch called it *yerw*, and not too far removed from either is the modern German name, *Garbe*.

Whether it was the species *millefolium* (Latin for "thousandleaf," for its fernlike foliage) or one of the other yarrows that was used by Achilles, we will never know for certain. However, *A. millefolium* is the one that is the traditional wound-healing species, the one whose leaves Mattioli described as being "like the wispy feathers of young birds." Throughout the millennia, throughout much of the world, until after the American Civil War, *A. millefolium* was as normal a part of battle gear as were the weapons that inflicted the wounds it helped heal.

Dioscorides had already named it soldier's woundwort; with an application of yarrow, he also prevented ulcers from becoming infected. To Romans, the weed was the *herba militaris*. Anglo-Saxons considered it a "wound herb," especially if the wounds were inflicted by iron. And during the Crusades, yarrow came to be called knight's balm and knight's milfoil. In the early seventeenth century, a poultice combining yarrow, southernwood, cumin seed, fenugreek, and dittany "bruised with black soap" was believed capable of drawing out any "splint, iron, thorne or stub."

The English name, carpenter's weed, and the French *herbe de Saint Joseph* are both founded on an old legend, wherein Saint Joseph injured himself while working at his carpenter's bench; yet when the Holy Child brought him a sprig of yarrow, the wound healed instantly.

Early immigrants introduced the plant to North America. As these settlers pressed across the continent, yarrow found new soils to settle in wherever it went. For nearly fifty years, from 1836 until 1882, yarrow was included in the *U.S. Pharmacopoeia*. Among American Indians, forty-six tribes, geographically as far apart as the Blackfoot and Micmac, Miami and Ute, came to rely on the plant's virtues for more than two dozen ailments. The Navaho know yarrow as the "life medicine." The Chippewa call it *a'djidamo'wano*, meaning "squirrel tail," perhaps in reference to the feathery leaves. The Shakers, those well-known specialists in herbal culture, attributed ten different properties to the weed, while English herbal tradition included yarrow among that small, select group of herbs known as "allheal." As such, it was a respected remedy even for melancholy, in the bleak, remote coastal regions of Scotland.

Yarrow seems really almost too good to be true; it does not appear to have any enemies, except perhaps itself. That is, the *millefolium* species excretes a toxin into the soil that gradually impedes and reduces the plant's own growth.

With so much versatility at its command, yarrow, not surprisingly, has also always been closely associated with witchcraft and matters magical. In many countries, it was used to ward off evil powers. In France, it became one of the herbs of Saint John, to be ceremoniously burned in the bonfires lit in honor of the saint, on Midsummer Eve. Yet, by others it was known as devil's nettle, or devil's or bad man's plaything, and ranked by some among the herbs formerly dedicated to the devil.

The Saxons wore charms stuffed with yarrow to protect them against highway robbers and other dangers, as well as against blindness. According to a nineteenth-century writer, one Elspeth Reoch was on trial for her life in Scotland during the 1650s, on the grounds that she had been supernaturally instructed to cure distemper with yarrow. Apparently, she did this "by resting on her right knee while pulling 'the herb callit malefour' betwixt her mid-finger and thumbe, and saying of, 'In nomen Patris, Filii, et Spiritus Sancti.'"

A verse of incantation, called "May Eve," used to be recited by witch cult followers in England, at the second of their four annual assemblies:

> There's a crying at my window and a hand upon my door,
> And a stir among the Yarrow that's fading on the floor,
> The voice cries at my window, the hand on my door beats on
> But if I heed and answer them, sure hand and voice are gone.

Although it may be true that all's fair in love and war, a local belief in some eastern counties of England, until as recently as early this century, took a rather masochistic turn. To divine the devotion of a lover, the seeker after this truth was required to tickle the inside of his or her own nostril while simultaneously reciting the following verse:

> Yarroway, yarroway, bear a white blow,
> If my love love me, my nose will bleed now.

There *must* have been other ways to find out.

By comparison, the following spell, also carried into this century, sounds much more encouraging. According to this ritual, a vision of the great love in one's life could be induced to appear in one's dreams by simply sewing an ounce of yarrow in a piece of flannel and placing it under the pillow at bedtime, while repeating this verse:

> Thou pretty herb of Venus' tree,
> Thy true name it is Yarrow;
> Now who my bosom friend must be,
> Pray tell thou me to-morrow.

In spite of its prevalence along roadsides, on the borders of fields, and in meadows, yarrow is somehow far less conspicuous than chicory or wild carrot. Yet its pungent aroma often lingers in the air around fresh-mown hay. In overall appearance, it is a delicate plant. It may be as short as 8 inches or as tall as 3 feet, erect or recumbent, single or forked. All of it is covered with fine white hairs, and it blooms from late May to November. The flowers are a creamy white, sometimes flushed with mauve, sometimes a warm crimson. They are not unlike daisies in form, and somewhat reminiscent of nested dolls. Seen from the inside out, so to speak, innermost are small groups of tiny, yellow-centered disk florets. Each group is surrounded by five petallike white ray flowers, with the result that each flattened, loose terminal cluster consists of many little flower heads made up of a few even smaller flowers.

Yarrow can be easily raised from seeds, or cuttings, and demands no special attention or care, although it does need full sun to do its best. Yarrow responds well to an occasional dusting of bonemeal, and it produces its most appealing fragrance in light, sandy soils. So that it does not gradually succumb to its own toxins, it should be thinned out and transplanted to another location every other year or so.

Yarrow may no longer be viewed as the medicinal panacea it once was, but even now, it is a force to be reckoned with in nature's own medicine chest. If, in the course of time, popular usage has credited yarrow with the powers to heal almost every ailment known to mankind, modern research justifies such zeal at least in part. The tannins in yarrow are astringent, its essential oils are antiseptic, and its action is hemostatic. It is cited as being exceptionally high in chlorophyll, as well as in protein and other nutrients that stimulate the appetite and purify the blood. It is also said to possess some aspirinlike derivatives.

And so, even by contemporary herbalist standards, yarrow's reputation remains untarnished in the treatment of wounds and festering sores, in anemia, intestinal and digestive disorders, and numerous other internal and external afflictions.

Yarrow is also used for cosmetic purposes and as a dye (it produces yellow and olive green hues), and even as an occasional substitute seasoning for nutmeg and cinnamon; it can be used in fresh or dried floral arrangements and as a long-blooming ornamental plant in the flower bed. To gardeners, it is a boon companion in the herb and vegetable beds. In fact, many gardeners call it the herb's herb: Its presence increases the essential oils of other herbs and improves their health; it attracts beneficial insects and wards off destructive ones, including the evil-spirited Japanese beetles.

The entire herb may be employed, fresh or dried, although the leaves and blossoms are considered best. Because the leaves tend to blacken soon after they are picked, they should be dried as quickly as possible. The whole weed is strongly aromatic and somewhat bitter in taste and odor, yet simultaneously refreshing, whether drunk as a tea or applied externally.

In medieval England, yarrow was drunk with "wine or good ale" to stop the "heartburning." The weed was also used to induce as well as to stanch a nosebleed. You may be puzzled why anybody would *want* a nosebleed, but it was believed to relieve headaches. The method required gently twirling a yarrow leaf in the nostrils until the desired result was achieved. To obtain the opposite result, John Parkinson wrote, "if it be put into the nose, assuredly it will stay the bleeding of it." Since dried yarrow was sometimes also snuffed, it is quite possible that this was done for the same purposes.

By describing the following experience of an acquaintance of his, John Gerard probably precipitated that young chap into a state of great embarrassment or even notoriety in the small world of a sixteenth-century university town. To relieve the "swelling of those secret parts," the young man "lightly bruised the leaves of common yarrow with Hog's grease, and applied it warm unto the privie parts, and thereby did divers times help himself and others of his fellows, when he was a student and a single man living in Cambridge."

"An ointment of the leaves cures wounds," wrote Culpeper, "and is good for inflammations." To stop hemorrhages and also as an excellent treatment for bleeding piles, he recommended "plentifully" drinking a strong decoction of the leaves boiled with white wine. "Equal parts of it, and of toad flax," he went on, "should be made into a poultice with pomatum [a perfumed unguent], and applied outwardly. This induces sleep, eases pain, and lessens the bleeding."

A German herbalist of the late Middle Ages, Jakob Theodor von Bergzabern, alias Tabernaemontanus, prescribed yarrow for patients suffering from listlessness or from a loss of thirst and appetite. Boil it in wine, he wrote, "strain it and drink sober every morning a warm cupful." Though prepared with water nowadays, yarrow tea is still prescribed as a digestive aid and tonic.

Linnaeus wrote that Norwegians used crushed fresh yarrow leaves to ease rheumatic pains. Calling the weed "field hop," the Swedes found an entirely different use for yarrow—in the production of a beer, which Linnaeus regarded as more intoxicating than beer made with hops. According to Mrs. Grieve, yarrow was put to similar use in some parts of Africa.

Among their numerous uses for yarrow, American Indians drank it as a tea, applied it as a poultice, chewed the fresh leaves, and used an infusion of it to wash infected eyes, to heal earaches, or to cleanse wounds and external infections. Like other peoples around the world throughout the weed's history, American Indians also deemed yarrow to be beneficial in treating various feminine complaints—and as an abortive. As regards the latter, research appears to have established that the plant contains some thujone, sufficient quantities of which could cause abortion. In fact, American Indians found yarrow beneficial not only for their physical health but for their spiritual well-being.

Germany's Father Kneipp considered an infusion of yarrow as an excellent diuretic and a valued treatment for lung ailments. For general debility, he pre-scribed 1 to 2 cups of infusion daily; for colds he recommended an inhalant, prepared by pouring boiling water over a small handful of the leaves and blossoms in a bowl. As a remedy for circulatory problems and hemorrhages, however, Kneipp regarded the freshly expressed juice as greatly superior to tea. He also preferred using the juice to treat badly healing wounds and tumors, as well as psoriasis.

Today's herbalists, in the United States and elsewhere, prescribe yarrow for many of the same ailments, and regard it as a general tonic that also benefits the entire nervous system. Some English practitioners of herbal medicine treat severe colds, the early stages of fevers, children's measles, and other eruptive diseases with cups of warm yarrow tea, to which honey and a little cayenne pepper, gingerroot, or Tabasco sauce have been added. They prepare an equally beneficial infusion with yarrow, elder flowers, and peppermint. With the honey-sweetened infusion, they treat kidney disorders, incontinence of urine, bronchitis, and arthritis. Unsweetened, both the infusion and decoction are traditionally respected preventives of baldness and are used in compresses for external wounds or bleeding piles. The fresh juice or a cooled, strong infusion, dropped into the eyes, wrote William Smith, takes away redness and a bloodshot appearance; chewing the fresh leaves eases toothache, he claimed.

To treat inflammation in any part of the body, an otherwise unidentified Dr. Box of England is said to have suggested taking freshly prepared cupful doses of a special tea. This tea required that 1 to 2 teaspoons herbal mixture composed of equal parts of yarrow, peppermint, and elder flowers be steeped in boiling water with a dash of cayenne pepper, strained, and slightly cooled. But my favorite part of his prescription is the accompanying admonition, spoken in the voice of a family doctor who still made house calls. This remedy, he wrote, "is of more value than all the orthodox drugs (ancient or modern) put together. Fast the patient, keep him or her well wrapped up warm in a well ventilated room and keep on giving frequent doses of the infusion hot until a full relaxation of the skin takes place and heavy perspiration ensues." That's telling them!

In France, where yarrow is viewed as a valuable decongestant and antispasmodic, an infusion of the herb taken internally was prescribed to remedy circulatory disorders, to ease menstrual and menopausal complaints, to restore liver function, and to soothe the pain of varicose veins.

For chronic mucous inflammations, catarrhal conditions, gastric, intestinal, and urinary disorders, German herbalists have traditionally recommended taking the freshly expressed juice of yarrow blossoms and leaves, combined with watercress, in daily doses of 2 to 4 tablespoons. For sore eyes, they suggested using an infusion both as tea, 2 cups daily, and as a gentle eyewash. For flatulence, headaches, and respiratory complaints, Maria Treben suggests taking 2 to 4 cups daily of an infusion steeped no longer than 30 seconds. For arthritis, dizziness, and constipation, she recommends a similar infusion.

A long-favored German treatment for numerous afflictions is the herbal bath. In the case of yarrow, this is considered particularly valuable for treating incontinence of urine and neuritis of the arms or legs. In fact, it is said to invigorate the entire body. For an herbal bath 3 to 4 ounces fresh herb are steeped overnight in 5 gallons cold water, brought to a boil the next day, strained, and added to warm bathwater.

For an aching back, both a tea and an herbal bath made with yarrow were particularly welcomed by European farmers, in the days when haying and harvesting were still done by hand and often on steep hillsides. But Abraham How, a creative and determined nineteenth-century English shopkeeper, developed what is surely the most dynamic treatment for back pain. All he did was to combine yarrow with brandy and gunpowder, plus some comfrey and borage. Alas, he didn't record whether to eat or sip this remedy—or to set a match to it!

Cosmetically, yarrow's astringent properties make it an excellent herb for a facial steam, particularly for oily skin. To combat oily hair, all that is considered necessary by some herbalists is a pot of yarrow tea—one cupful to be drunk every day, half the remainder to be used for rinsing the hair after shampooing, the other half to be used as a skin lotion. Daily application of this lotion is said to improve the complexion and to rid it of pimples.

If blackheads are a problem—and when are they not?—the only way to eliminate them permanently, according to a Dr. Anna Kingsford, is to apply an herbal milk. This is prepared by soaking 4 tablespoons chopped yarrow in 4 ounces cold milk, and refrigerating the mixture overnight. The herbal milk should be heated to lukewarm and strained. After first washing with water as warm as possible, one should "bathe the face (with a sponge) for 10 minutes in tepid milk," Dr. Kingsford recommended. For added benefit, English herbalists suggest also drinking a cup of yarrow infusion every day.

Even corns and calluses can be eased with yarrow—as can aching, tired, swollen feet. For a wonderfully refreshing oil, I squeeze the juice of half a lemon into my soup, salad, or dessert, then turn the peel inside out, place it in a cup or small bowl, fill it with sunflower oil, and set it in a warm place overnight (on a radiator, for instance). For my fairly frequent footbaths, I infuse 2 to 3 handfuls fresh yarrow in 1 quart boiling water for 1 hour, add the infusion to the footbath water, and soak my feet for 20 to 30 minutes. After drying them, I thoroughly massage them with the lemon-peel oil. What a treat!

Of course, yarrow can also be eaten, although it never gained great popularity. On its own, eaten fresh, I find its flavor too pungent, rather bitter. Still, a small quantity of finely chopped leaves adds an interesting flavor to a potato salad, or to a mixed green or vegetable salad, for example. But the use for which it is perhaps best known, even today, is as one of the fresh ingredients in what German folk tradition calls the healthiest spring herbal soup. As soon as the fresh green growth begins to sprout every year, women scour the fields for a wide variety of wild plants. With these they prepare teas, salads, and soups, to rid their bodies of toxins and to purify their blood. Although the ingredients vary, depending on their availability, the following is a typical recipe that takes advantage of some varied spring growth, and serves 2 to 3 persons.

Three tablespoons mixed herbs (yarrow, dandelion, chickweed, stinging nettle, sorrel, ground ivy, purslane) are washed and chopped, then sautéed in 1½ ounces butter until wilted. Next, 1½ ounces flour are stirred in until pale golden brown, to which 2 pints chicken stock or water are added all at once, rapidly blended with a whisk, brought to a boil, seasoned with a little salt, and simmered for 5 minutes. Lastly, a beaten egg is whisked into the soup just before serving.

Bon appétit! Or should I say, *À votre santé?*

Agropyron repens

Agropyron repens
(C O U C H   G R A S S)

VITAL STATISTICS

Perennial

COMMON NAMES	*(Triticum repens)* Couch Grass, Witch Grass, Quick Grass, Twitch Grass, Dog Grass, Quack Grass, Dutch Grass, Quelch
USES	Culinary, medicinal
PARTS USED	Underground rhizomes
HEIGHT	1 to 3 feet
FLOWER	Tiny, purplish stamens in eight or more spikelets atop jointed, hollow stems; similar to rye; May to September
LEAVES	Grasslike, flat, slender, sharp; rough on upper surface
ROOT TYPE	Creeping subterranean rhizomes, smooth, succulent, hollow, yellowish white; tiny rootlets at nodes
HABITAT	Gardens, fields, waste places, seashores
PROPAGATION	Creeping rootstock; seeds
CONTROL	Patient, repeated weeding; frequent cultivation; one or two plantings of millet, soy beans, or rye crops

I t is no wonder that couch grass is one of the most detested weeds in the life of a gardener or farmer. Divide and conquer seems to be its motto, and once it has gained a foothold, it is almost ineradicable. Left free to roam where it chooses, it spreads at an alarming rate, its jointed underground rhizomes branching in all directions. Wherever the rhizomes travel, each of their countless nodes is capable of sending up a new shoot of grass. And each of these forms its own rhizomic system, interlaced with the others into a dense matting that ruthlessly pushes its way over, under, around, and among the roots of any other plants in its way.

Of course, all of these nasty habits can be considered virtues, if preventing soil erosion is the objective. However, in a garden, even when couch grass is meticulously weeded—and it is virtually impossible to do so without breaking the roots or stems—if only the tiniest scrap remains in the soil, each piece develops endless new plants with unnerving zeal and success.

In my earliest encounter with couch grass, I spent one entire summer painstakingly removing what I thought was every particle of the weed from a particular flower bed. "The stalks are jointed like corn," Nicholas Culpeper's description had told me, "with the like leaves on them, and a large spiked head, with a long husk in them, and hard rough seeds in them." His paragraph ended, "If you know it not by this description, watch the dogs when they are sick, and they will quickly lead you to it." I chuckled, having misinterpreted this last comment. It somehow reflected the loathing I had come to feel for this weed. Nevertheless, day after day, I carefully traced and lifted each rhizome, each shoot, each root. I even dug up every single garden plant in that bed, freeing the roots from the stranglehold of the intertwining couch grass rhizomes.

By early autumn I felt justly proud of that bed. It looked wonderfully groomed. Naturally, I could hardly wait to add new flowers to it the following spring. Ha! When spring came, *Agropyron repens* let me know in no uncertain terms that not for nothing has it survived through the ages. Only then did I begin to wonder whether perhaps it served some purpose other than to raise a gardener's blood pressure.

15

In Culpeper's time, the plant's Latin name was *Triticum repens*, or "creeping wheat," and, for its content of the carbohydrate triticin, *triticum* is the name by which couch grass is known in modern pharmacology. Linnaeus and subsequently the eighteenth-century French explorer Ambroise Beauvois renamed it *Agropyron repens*—which is how the plant is usually listed—from the Greek *agros* (field) and *puros* (wheat). However, both Dioscorides and Pliny wrote of a grass having creeping roots like couch grass, which they knew as *Agrostis* and *Gramen*. In fact, both these men claimed, as have most other physicians and herbalists since then, that a decoction made from this grass serves as a remedy for various urinary complaints.

There is no doubt that the plant we know as *A. repens* enjoys a respected tradition in folk medicine, particularly in Europe, where the dried rhizomes, freed of the hairy rootlets and cut into short lengths, are available commercially. In Italy, they used to be—perhaps still are—sold fresh in the markets. In Germany, the rhizomes have long been prescribed to induce sweating and as herbal treatments of urinary and rheumatic ailments and chronic diseases of the liver and spleen.

And in France, the prominent eighteenth-century chemist and physician Count de Fourcroy observed that the young stems and leaves of couch grass "actively yet gently prevent the formation of gall-stones." Even in modern France the plant is prescribed, in the form of a tea, as both a sudorific and a demulcent. French herbalists also recommend it in daily doses of 3 to 5 cupfuls for rheumatism, gout, kidney and liver disorders, jaundice, bladder stones, chronic catarrh of the respiratory tract, gastrointestinal inflammations, and skin eruptions.

A decoction of couch grass should always be prepared in two waters, to eliminate the bitter taste of the rhizomes. That is, 1 ounce of fresh or dried rhizomes is soaked in cold water overnight, then boiled for one minute, strained, and the water discarded. The rhizomes are then added to 2½ pints hot water and boiled until the liquid is reduced to 2 pints. To mask the remaining bitterness, many people add a little honey and a lemon slice.

Although couch grass seems to have been relegated to the ranks of useless weeddom in modern-day England, both Gerard and Culpeper agreed on its beneficence. "Although that Couchgrasse be an unwelcome guest to fields and gardens," wrote Gerard, "yet his physicke virtues do recompense those hurts; for it openeth the stoppings of the liver and reins without any manifest heat." For his part, Culpeper considered it "the most medicinal of all the quick grasses." He recommended that "the way of use is to bruise the roots, and having well boiled them in white wine," to drink the decoction. He described this as "very safe; 'tis a remedy against all diseases coming of stopping, and such are half those that are incident to the body of man." In fact, Culpeper went so far as to say that, "although a gardener be of another opinion, yet a physician holds an acre of them [couch grass] to be worth five acres of carrots twice told over."

Those are fighting words! It is doubtful that the sentiments of modern herbalists would go quite so far. Nevertheless, the entire plant is described as a diuretic and a demulcent; its high content of mucilage is considered particularly palliative to the mucous membranes. Its rhizomes are known to be rich in nutrients—in minerals and vitamin B_1—especially when freshly harvested in spring or autumn.

Although some scientists question the use of couch-grass tea, Europeans have long prepared a decoction of the weed as a popular spring drink for purifying the blood. To do so, they simmer 1 ounce of the rhizomes in 1 pint of water for 15 minutes, then strain the liquid, drinking it in doses of one small cupful for 2 or 3 days.

To prepare the *juice*, German herbalists recommend crushing the fresh rhizomes in a mortar, then expressing the juice from them through a cheesecloth, and taking 1 teaspoon in a glass of water or light wine. An old French recipe recommends extracting the juices from equal quantities of couch-grass leaves and stems, dandelion leaves, and plantain leaves, and taking them in 4-ounce doses before breakfast for 3 or 4 days. A modern French physician, Dr. Cazin, is convinced that people would "lead a sober life, and feel all the better for it," if they drank such herb juices after "the dissipated life they led in winter, suffering from gout, gallstones, haemorrhoids [sic], scabby skin conditions, acne, etc."

In spite of the fact that cattle, goats, and sheep thrive on couch grass, and dogs seek it out for its emetic properties, when they are sick—hence the name dog grass—the plant has never gained a reputation as a food. Only in times of dire need, we are told, have the dried and ground rhizomes played a role in the human diet—as a coffee substitute and even as an emergency flour for making bread. Yet it is interesting to note that couch grass is considered the nearest relative to wheat, as is already indicated by both its botanic names. Therefore, it is all the more interesting to learn that this ancient "creeping field wheat" has so far defied scientific experiments to transform its weedy vitality into a productive, *upright* property. For the present, at least, Nature seems inclined to keep the explanation for this truculence locked in its own laboratory.

Arctium lappa

Arctium lappa

(G R E A T   B U R D O C K)

V I T A L   S T A T I S T I C S

Biennial

COMMON NAMES	Great Burdock, Burdock, Burrs, Beggar's Buttons, Fox's Clote, Hardock, Cockle-Buttons, Stick-Buttons, Bardane, Burrseed, Clotbur, Harebur, Hare-Lock, Turkey-Bur, Gobo, Personata, Happy-Major, Lappa, Thorny Burr, Love Leaves, *Philanthropium*, Gypsy's Rhubarb, Pig's Rhubarb, Snake Rhubarb, Dock, Horseburr
USES	Culinary, medicinal, cosmetic
PARTS USED	Whole plant
HEIGHT	To 8 feet
FLOWER	Purplish red, in burrlike heads; July to October
LEAVES	Ovate, to 20 inches long, grayish downy undersides
ROOT TYPE	Taproot, to 3 feet
HABITAT	Roadsides, fields, waste places, stream banks
PROPAGATION	Seeds (burrs), roots
CONTROL	Digging up roots, deadheading

It is not impossible that the recorded history of burdock, though relatively brief at present, will one day be traced back thousands of years to the twentieth century as the inspiration for Velcro. Anyone who has ever roamed alone, or with children and pets, through late-summer fields, beside streams, or along roadsides or woods will know whereof I speak! It's those wretched burrs, adhering as they do to everything they touch, on man or beast. As Celia tells Rosalind in Shakespeare's *As You Like It*, "if we walk not in the trodden paths, our very petticoats will catch them."

Burdock's botanic name, derived from the Latin *arctos* for "bear" and for "north," conveys both its formidable size and its generally preferred habitats—north of the Equator, throughout Europe, Asia, and North America. The specific Latin name, *lappa*, refers to the burrs. Mrs. Grieve tells us that "an old English name for the Burdock was 'Herrif,' 'Aireve,' or 'Airup,' from the Anglo-Saxon *hoeg*, a hedge, and *reafe*, a robber—or from the Anglo-Saxon verb *reafian*, to seize."

Most of the weed's common names are clearly local or regional, although for the name *happy-major* no likely source immediately leaps to mind. Only Shakespeare offers a modest scholarly challenge, in *King Lear*, when he has Cordelia say, "Crown'd with rank fumiter and furrow weeds, with harlocks, hemlock, nettles, cuckoo-flowers. . . ." Did Shakespeare mean burdock when he wrote "harlock"? Scholars are undecided. Did Shakespeare write "harlock" at all, or was that *l* an ill-written or ill-read *d*, in the old script typical of the Bard's time? *Webster's* defines only *hardock* (also spelled *hordock*) as "an unidentified plant mentioned by Shakespeare, perhaps burdock." Certainly, the first syllable, *har*, Anglo-Saxon for "white or gray" (as in *hoar*frost), supports this assumption; a quick glance at the light gray covering of fine down on the underside of burdock leaves will make that apparent. *Bardane* is merely French for "burdock," while *personata* suggests Latin for "impersonation," and *philanthropium* is undoubtedly the latinized version of the Greek *philanthropos*, "loving mankind."

Whatever names burdock goes by, even without its burrs it demands our attention. It is a majestic plant, with its flower stalk often more than 6 feet tall, topped with heavily armored, frail-looking, clustered purple flowers in summer, and with basal leaves that are not infrequently long and wide enough to hold a baby. Although the weed grows in almost any soil, it does best in rich, light, well-drained soil. Not surprisingly, therefore, it turns up all too often in our flower beds. And that spells *trouble*, because digging up a burdock can be a nightmare: Its taproot may reach a depth of as much as 3 *feet!* This means you may have to ram down a posthole digger beside it to help pry the root loose.

Who knows, such aggravation alone—on the if-you-can't-lick-'em-join-'em principle—may be the reason why burdock has been actively cultivated in many countries throughout much of its centuries-old history. You may wish to cultivate it as well once you know about the many uses to which it can be put, especially that taproot. If you do, plant the seeds in deeply compost-enriched, loose, and *mounded* soil, and use only first-year roots. Remember that burdock is a biennial; that is, the seeds that were broadcast when last year's bristly flower clumps turned into those infamous burrs will develop into new burdock plantlets this spring. Their roots will be ready for digging either this autumn or next spring. Even these, however, will resist extraction to some extent.

All parts of the weed can be used, either medicinally or gastronomically—or both. The plant is rich in inulin, a form of starch that is the source of most of burdock's curative powers; it also provides an abundance of iron and mucilage, some tannic acid, volatile (essential) oils, sugar, protein, and vitamins A, B, and C. As Culpeper noted already three centuries ago, "the root beaten with a little salt, and laid on the place, suddenly easeth the pain thereof, and helpeth those that are bit by a mad dog." He recommended that "the juice of the leaves, or rather the roots themselves, given to drink with old wine, doth wonderfully help the biting of any serpents," and that "the seed being drunk in wine forty days together, doth wonderfully help the sciatica: the leaves bruised with the white of an egg and applied to any place burnt with fire [fever], taketh out the fire, gives sudden ease, and heals it up afterwards: the decoction of them fomented on any fretting sore or canker, stayeth the corroding quality, which must be afterwards anointed with an ointment made of the same liquor, hog's grease, nitre, and vinegar boiled together."

Many of burdock's traditional uses are still deemed effective today. In fact,

burdock has always been considered one of the most valuable of all blood puri-
fiers, boosting kidney function in ridding the blood of harmful acids. Most often
it is prescribed in the form of a decoction made from the fresh or dried roots or
from the dried seeds, or in the form of a tea made from the fresh or dried leaves,
in doses of a small cupful three or four times daily. Whereas burdock roots have
a sweetish flavor, the leaves taste slightly bitter, and are generally sweetened with
honey. Dried burdock, available in tea blends, is highly regarded as both a valuable tonic and a remedy for chronic indigestion. My sister Anne remembers one
of our uncles regularly sending her as a child to buy dried burdock roots at the
local chemist's in London.

A wineglassful of a decoction made of the seeds, taken three times daily, is
suggested by many herbalists not only for relaxing the body but also for improving the skin. Because of their oily nature, burdock seeds are believed to be capable
of simultaneously cleansing the sebaceous and sweat-inducing glands and of restoring the skin's natural smoothness. Because it has been found to induce sweating, a cup of burdock tea drunk before taking a hot bath is believed to cool the
body and clear fevers. On the same principle, Jean Palaiseul claims it to be little
short of a miracle cure for measles. He boils 25 to 30 grams of fresh root in half
a liter of water, strains them, and adds some honey. He suggests, "Administer the
preparation in dessertspoonfuls every five minutes. Within a few hours, eruption
is completed and, keeping the young patient warm, he will recover in three or four
days." Drinking an unsweetened decoction of the fresh root has also been credited
with lowering the blood-sugar level.

Because burdock has long been considered an excellent remedy for all skin
diseases, including acne, and of great benefit in the treatment of arthritis, sciatica,
lumbago, and gout, burdock teas have often been accompanied by external applications of a poultice. "Boil the leaves in urine and bran, until the liquid is almost
gone; apply the sodden remains to the affected area," suggests one rather extreme
recipe, as quoted in *Weiner's Herbal*. But the simple truth is that, minus at least one
of the above ingredients, bruised or crushed burdock leaves have been relied on
for centuries by various peoples around the world for relief of gouty swellings,
tumors, skin inflammations, and sore, swollen feet. A poultice of burdock leaves,
or the pulp of fresh root, applied to boils, bruises, carbuncles, infections, sties, and

canker sores, is thought to be particularly effective if a decoction of the root is drunk as well. The fresh leaves of burdock, when the downy side is applied to the skin between the shoulder blades, are traditionally reputed to soothe diseases of the respiratory tract. In the Middle Ages, burdock leaves pounded and macerated in wine were believed to effect a remedy for leprosy. Citing the use of burdock in such chronic skin diseases as psoriasis and eczema, the United States Dispensatory of 1888 stated:

> To prove effectual, its administration must be long continued. A pint may be given daily of the decoction, made by boiling two ounces of the root in three pints of water down to two pints. The fluid extract and syrup have also been prepared. The fresh leaves have been employed both externally and internally in cutaneous eruptions [skin breakouts] and ulcerations.

From personal experience I know that sprains and bruises can be relieved with the application of crushed fresh burdock leaves. This treatment also almost instantly relieves the itch and pain of a mosquito bite or other insect sting. Or, if I have sunned myself to a crisp, compresses dipped in a cooled infusion of burdock leaves soon soothe my burning flesh—and also keep the skin elastic. A treat my feet truly relish, especially if I've made heavy demands on them all day, is burdock oil. To prepare this, I loosely pack coarsely chopped burdock leaves in a pint jar containing 1 cup of almond or sunflower oil. I set the closed jar in the sun for three weeks, shaking it daily. Then I pour the contents into an enameled pan and lightly sauté the leaves over low heat, strain them, and bottle the oil. A little of this massaged into the feet whenever they feel weary soon perks them up. Lastly, many people have combated thinning hair by applying one or another of the following three remedies: (1) repeatedly applying compresses soaked in a strong decoction of burdock root; (2) daily massaging the scalp with a lotion made from finely chopped burdock and stinging-nettle roots soaked in rum for a week; or (3) massaging the scalp with a lotion containing equal parts of a decoction of burdock root and wine or apple-cider vinegar.

That brings us to food. My own favorite common name for this weed comes from Japan, where it is called *gobo*. Japanese truck farmers in Hawaii are said to sell the cultivated weed as a vegetable whose sliced roots are used in sukiyaki. In fact, many people believe that eating gobo not only gives strength and endurance but also acts as an aphrodisiac.

Whether or not this last is true, as tough as burdock is to extirpate from the garden, it can be a most delicate, versatile addition to the dinner table, its flavor a blend of celery and potato, with perhaps a soupçon of cucumber. It is probably most frequently employed in Japanese cookery, but in Western dishes, too, its mild flavor has often made it a delightful substitute for asparagus and a welcome ingredient in salads, vegetable dishes, soups, stews, pickles, and relishes. The root may even be candied, a practice that was already well known in Culpeper's time.

All parts of burdock are considered best when they are collected from first-year plants, before these go to seed. The *very* young leaf tips can be eaten raw in early spring, added to green or mixed salads. However, because of the leaves' somewhat bitter taste, some people prefer to cook them. This is done by boiling the leaf tips for 5 minutes in each of 2 changes of salted water to cover, to which a pinch of baking soda has been added. In this way, both the bitterness and the natural coarseness of the leaves can be mitigated. Once cooked, the leaf tips may be served up as a warm or cold salad, alone or together with other salad ingredients, dressed with a simple vinaigrette. As a vegetable, the spring leaves of burdock may be steamed or sautéed, served plain or creamed like spinach, or with butter or a hollandaise sauce.

If it is to be prepared as a food, burdock root, probably the best-known part of this weed, should always be peeled. This exposes the delicious core, whose diameter is roughly half an inch. Once peeled, burdock root may be sliced lengthwise or crosswise, or coarsely chopped, then added to salted boiling water to cover, to which a pinch of baking soda has been added. Cooked at a slow boil for 15 to 20 minutes and drained, the root is added to fresh boiling salted water to cover, and simmered for another 7 to 10 minutes. Such double boiling of burdock root, with a change of water after the first boiling, is often recommended. (This second pot of water can become the base for a vegetable soup, or used in lieu of water in the making of chicken broth.) The cooked root can be served with lightly browned butter and a sprinkling of pepper and parsley, or prepared like scalloped potatoes, in a baked casserole. Sliced thinly and twice soaked in cold water with a pinch of soda for 30 minutes each time, burdock root can also be added to soups and stews at serving time. Or it can be sautéed with onion in a stir-fry that includes green peas and julienne-sliced carrots, seasoned with freshly shredded ginger, minced garlic, soy sauce, and lemon juice.

For many individuals the tastiest part of the great burdock reposes in its flower stalks. These may be already quite tall and more than an inch thick, but the crucial factor is that they are gathered during the formative stage of the flower heads. It is essential to peel off the thick green rind, revealing the core, the tender white pith. This can be cut into 6-inch lengths and prepared like asparagus, to be either steamed or boiled in two changes of salted water with a pinch of baking soda. The delicately flavored pseudoasparagus can then be served with a little butter or with hollandaise or béchamel sauce. Or dot them with butter flecks, sprinkle with cheddar cheese, and bake until the cheese is golden brown. Another use is to cut them into mere 1-inch lengths and add to chicken salad, together with halved orange slices and chopped walnuts; or add to vegetable soups, meat stews, omelets, and soufflés. They can even be served as cream of "asparagus" soup.

For teas, the root has always been used, either fresh or dried. The drying process can be accelerated if the roots are first sliced fairly thinly, then spread out on cotton cloths or brown paper in a warm and dry location. The leaves may be dried in the same manner.

Both the sliced roots and stalks may be prepared as a tasty confection. They are first boiled gently for 15 to 20 minutes in plenty of water, to which a pinch of baking soda has been added, then drained thoroughly and set aside. Two parts sugar to one part water are brought to a rolling boil together with the juice and grated peel of one lemon and a handful of finely chopped fresh sweet cicely leaves. Next, the burdock pieces are added and boiled in the syrup for 20 to 30 minutes, drained, and cooled. If desired, the pieces can be rolled in sugar.

With so many beneficial options before you now, in sickness and in health, I hope that you will come to regard burdock as a blessing instead of a pest the next time you pry its burrs off your clothes or its roots from among your perennials.

Berberis vulgaris

Berberis vulgaris

(B A R B E R R Y)

Perennial Shrub

COMMON NAMES	Barberry, Common Barberry, European Barberry, Jaundice Berry, Pepperidge Bush, Berbery
USES	Culinary, medicinal, commercial
PARTS USED	Stem bark, root bark, berries, leaves
HEIGHT	5 to 10 feet, upright, bushy, deciduous; woody, smooth stems, grayish bark; yellow inner bark and wood
FLOWER	Small, yellow, short stalked, hanging clusters; May to June
FRUIT	Bright red, tart, juiceless berries; August to spring
LEAVES	Grouped, ovate, spiny toothed, green above, grayish underneath; some leaves reduced to three-forked spines
ROOT TYPE	Fleshy, deep, branched; brownish yellow
HABITAT	Pastures, hedgerows, thickets, waste places, underbrush
PROPAGATION	Seeds, cuttings, rooted suckers
CONTROL	Weeding young seedlings; digging up older plants

The first legislation against common barberry was passed in 1670. Extreme as such a measure may seem, farmers claimed that even a single barberry bush could blight an entire field of wheat. Although such an "absurd" notion has met with a good deal of derision over the centuries, farmers have kept routing the shrub from their fields almost as fast as it can multiply. They know what they know, and nothing can change their minds.

The truth of this baffling matter did not fully emerge until early this century, when botanists learned that, in a sense, farmers had been right all along. It seems that the common barberry and several other barberry species are alternate hosts for a serious wheat rust, or wheat mildew. This means that barberry is a plant on which wheat rust depends for half of its life cycle, depending on a completely unrelated other plant to complete the cycle. In this instance, the unrelated other is wheat, which, in turn, reinfects the barberry. Consequently, in most areas, breaking the cycle by removing one of the hosts from the scene is a simple remedy. In cereal-growing regions, however, the entry of suspect barberries is simply forbidden—by state law.

The genus *Berberis* includes more than 400 species of deciduous and evergreen shrubs, spread mostly throughout the temperate regions of the Northern Hemisphere. More than fifty species are cultivated in the United States. Their brush with the law notwithstanding, all barberry species are regarded as horticulturally desirable, even the most prevalent of them, the common or European barberry.

Yet, in spite of its historic notoriety, surprisingly many people are unfamiliar with this plant, thereby depriving themselves of much pleasure throughout the seasons. Common barberry is that rare weed that instantly responds with grace to even the slightest grooming, and assumes its role as ornament with ease. In fact, few ornamental shrubs are as agreeable as barberry when it comes to being molded into new shapes by humans—square, squat, globular, pyramidal, or anchored to the ground as a topiary bird in flight, barberry retains its identity. On the other hand, left untended and ignored, it gradually becomes an unsightly, disheveled weed, hugging the base of a tree, sprawled against a house, or propped up by a neglected hedge.

Nevertheless, as weed or ornament, barberry has always remained true to its calling as a medicinal plant and a food. There is little doubt that the shrub is best known by those who have had personal hands-on experience with its stockpile of needle-sharp thorns. Exactly why Nature decided to make it so difficult (and painful) to gain access to the plant's several valuable properties, chief among these a yellow substance called *berberine*, is not known.

Scrape away any portion of a barberry's gray stem bark, or expose the root, and a deep, nearly orange-juice-yellow inner bark is revealed. Very likely interpreting the yellow color as a divine sign, the physicians of long, long ago found barberry good for treating jaundice, a condition that causes the skin to turn yellowish. Such an association was the basis of the once popular doctrine of signatures. According to this theory, a plant's dominant characteristic, appearance, or shape was believed to indicate the type of illness or injury it would cure.

This theory has not entirely withstood the test of modern science's analytical capabilities, but the barberry's yellow "signature" (nowadays better known as berberine) does have the grudging acknowledgment of researchers for improving liver functions. In small doses, studies have shown, berberine has stimulant effects on the heart; in large doses, the effects are depressant. Berberine is also considered tonic, purgative, antiseptic, and astringent, and said to contain sedative and anticonvulsant properties.

The bright red fruit of barberry is rich in vitamin C and pectin. To ward off pestilential fevers and agues, ancient Egyptians drank a diluted syrup made of the fruit of barberry and fennel seed. The shrub's antiseptic, antibacterial properties very probably made this a most effective remedy. Half a world away, American Indians knew that a decoction made from barberry's root bark and taken internally served to stimulate their appetite and to restore their strength after suffering illness, or when they were recovering from wounds. Applied externally, as a lotion or wash, barberry was used to heal skin infections and injuries.

Because the shrub's leaves have astringent properties similar to those of its berries, they are sometimes used for the same purposes. Not surprisingly, therefore, John Gerard recommended the use of barberry leaves, probably as a digestive aid, "to season meat with and instead of salad." For the same purpose, German herbalists today still prescribe a tea made from the leaves. For his part, Nicholas Culpeper stated that "the berries are as good as the bark, and more pleasing; they get a man a good stomach to his victuals." Presumably, they do the same for women and children.

Culpeper appears to have had strong feelings about barberry. Sounding like John Bull himself, he asserted that "Mars owns the shrub, and presents it to the use of my countrymen to purge their bodies of choler." He recommended boiling the inner rind in white wine and drinking 4 ounces of this decoction every morning to rid the body of sores and ringworm, jaundice, bile, and heat of the liver.

However, even in modern usage, both the stem bark and the root bark of barberry are used to improve the functions of the liver, gallbladder, spleen, and pancreas; they are considered beneficial for jaundice, general debility, and biliousness; they are said to regulate disorders of the digestion, and to act as a mild purgative, as an effective antiseptic, antidiarrheal agent, *and* laxative. They have also been used to treat edema and arthritis. In Germany, the root bark alone is employed to lower high blood pressure and to regulate blood circulation. In the United States, pharmaceutical eye preparations use berberine in eyedrops and eyewashes—just as old-time herbalists did.

The stem and root barks possess the same constituents and are used interchangeably; their taste is somewhat bitter. The bark pieces described below can be used fresh or dried, whole or powdered. They are generally prepared as a decoction (1 ounce bark pieces soaked in 1 quart cold water for 20 minutes; slowly brought to boil; immediately covered, removed from heat, infused 20 minutes; strained), suggested in doses of 1 small cupful three times daily, after meals. If the bark is dried and powdered, the following is suggested: ½ teaspoon bark to 1 cup boiling water, taken three times daily.

Once the thorns are removed, the stem bark can be gathered by shaving or peeling it off; because of the numerous nodes from the thorns, the portions are small and irregular. Externally, the pieces are grayish; internally, they are a rich shade of ocher. They can be dried in a well-ventilated, warm location, or in an oven (door ajar) set no higher than 200 degrees Fahrenheit. When the pieces are uniformly dry, they can be stored in tightly closed containers. Except for the thorns, the method for gathering the root bark is identical.

A tea made from the crushed fresh or dried berries (1 to 2 ounces soaked 2 hours in 1 quart cold water, boiled 1 minute; covered, cooled until lukewarm), sweetened with honey, and drunk in 4-ounce doses, is described not only as a pleasant, slightly acid drink for feverish patients, but also as useful for treating diarrhea. Either sweetened or unsweetened, this decoction is considered an excellent astringent gargle for sore throats.

With such a staunch reputation as a medicinal plant, it is surprising to learn that barberry was originally cultivated for its scarlet fruit. The berries were, and still often are, made into or added to jams, jellies, stewed fruit, chutneys, lozenges, and syrups. The city of Rouen, in northern France, used to be famous for what its restaurants and hotels so elegantly offered as *confitures d'épine vinette*, which translated into English as "common barberry jam" loses a soupçon of its charm. In some parts of England, barberry may still be sometimes referred to as piperidge or pepperidge bush. This name is said to have originated with the French words *pépin*, a pip, and *rouge*, red (and was possibly introduced by the Huguenots).

More than a century ago, the renowned Mrs. Beeton, who brought the traditions of English cookery to the attention of the world, suggested still other culinary uses for the fruit of this shrub: "Barberries are also used as a dry sweetmeat, and in sugar-plums or comfits; are pickled with vinegar. . . . The berries arranged on bunches of nice curled parsley, make an exceedingly pretty garnish for supper-dishes, particularly for white meats, like boiled fowl à la Bechamel; the three colours, scarlet, green and white contrasting so well, and producing a very good effect."

The fruit of barberry tastes not unlike rosehips or cranberries, perhaps a little more tart. The individual fresh berries can be used for uncommon cake or pie decorations; candied, they can be offered in place of after-dinner mints. A handful of them in soups and stews adds both delicate color and the slight sharpness that would normally come from lemon juice. A mixture made from equal parts barberry fruit and honey cooked down to a syrupy consistency can be stored until needed to soothe sore throats; or, diluted, the syrup can also serve as a gargle. For a cooling summertime drink, make a strong infusion with crushed barberry fruit, strain, and let it cool, then add it to your favorite fruit juice or soda.

Crushed and made into jelly, barberry makes a novel companion to toast, pancakes, muffins, waffles, or hot cereal (4 cups barberries boiled until soft in 3 cups water, strained through a jelly bag; equal parts juice and sugar or honey boiled rapidly and stirred until a drop on cold china forms a hard film). Barberry jelly made with apple cider (instead of water) and sugar, seasoned with sprigs of fresh peppermint, becomes a conversation piece at table, and can be eaten with cold cuts and cheeses, with potato or pasta dishes, hot or cold, or served with lamb, pork, venison, beef, or fowl.

For a simple and most unusual dessert for six to eight persons, I bring to a boil 8 ounces washed berries in 8 ounces water over medium heat. Stirring constantly, I lightly crush the fruit while adding 8 ounces sugar, return to boil, lower heat, and stir until fruit is soft (about 5 minutes). I purée this at once in a blender, and immediately stir in 1 envelope unflavored gelatin until completely dissolved. Then I fold in 1 pint vanilla ice cream until smooth, plus 1 ounce Grand Marnier or Cointreau (optional), and refrigerate until set, about 1 hour. Served with vanilla wafers, this is a tangy, refreshing change of pace.

By boiling the roots in lye, a rich yellow dye can be extracted to color wools, linen, and cotton; in Poland, it was not uncommon for people to use barberry bark to dye their leather or to polish it to a lustrous tawny finish. Similarly, they used barberry to stain certain woods, which they then used in marquetry work. Only Culpeper alone, that inveterate herbal enthusiast and invaluable source of information, attributes even a cosmetic usefulness to barberry; namely, he writes, "the hair washed with the lye made of ashes of the [barberry] tree and water, will make it turn yellow, viz. of Mars own color."

It is, of course, quite possible that you have no interest in using any part of this plant for any purpose whatever. Even then, however, if you happen to have a barberry shrub nearby, you will find it near impossible to ignore it through the changing seasons. In spring, it is swathed in a delicate mist of new greenery that is followed by the almost pointillist art of its tiny yellow blossoms; its summer coat is a rich, muted green that gradually assumes the tints and tones of a fiery autumn sunset. And then, their passion spent, the colors of the leaves withdraw to somber hues of purple and near-brown, and begin to fall. In a final burst of magic, barberry now fully reveals its mass of scarlet berries, to light the way into winter. By spring, the berries have shriveled and dropped; the cycle of colors can begin again.

Brassica nigra

Brassica nigra
(B L A C K   M U S T A R D)

VITAL STATISTICS

Annual

COMMON NAMES	Black Mustard, Mustard
USES	Culinary, medicinal, commercial
PARTS USED	Seeds, young leaves, flower buds
HEIGHT	To 6 feet, average 2 to 3 feet
FLOWER	Yellow, four petaled, in terminal clusters, May to October; tiny brownish black seeds in long slender pods, June to October
LEAVES	Various shapes, lower leaves broad, pinnate, deeply lobed, coarsely toothed; become smaller, narrower toward top
ROOT TYPE	Branched, fibrous
HABITAT	Waste places, roadsides, fields
PROPAGATION	Seeds
CONTROL	Frequent cultivation, weeding

Condiment, ancient drug, nutritious food, insect control, modern medicine, irritant, soil sweetener—above all, mustard is an aggressive weed! And yet it is so much more. It grows everywhere and anywhere. In spring, it sets whole fields and roadsides alight with its bright yellow blossoms shaped somewhat like miniature Maltese crosses. It is native to Europe, Asia Minor, northern Africa, India, China, and the Americas. In many countries, it is cultivated for its seeds and its oil, as well as for its greens.

Of the numerous plants called mustard, two have been cultivated for perhaps more than 4,000 years. They are the black mustard, *Brassica nigra* (sometimes *Sinapis nigra*), and the white *B. hirta* (sometimes *Sinapis alba*). Of these two, *B. nigra* is believed to have been the variety mentioned in the Bible and used since ancient times.

All the mustards belong to the Cruciferae, also known as the Brassicaceae. This vast family contains about 350 genera and some 3,000 species of plants distributed throughout the temperate and frigid zones of the world, particularly in the Northern Hemisphere. Scores of them are garden plants and weeds. Most of the plants earned their membership in the clan by having four-petaled flowers in the shape of a cruciform. Of them all, it is probably safe to say that the genus *Brassica* is most widely known. In addition to several of the mustards, it contains such stalwart friends of the human diet as cabbage, broccoli, cauliflower, brussels sprouts, kohlrabi, kale, turnip, collards, pakchoi, and rutabaga.

What may well be the earliest specific mention of mustard appears in that remarkable ancient record, the Code of Hammurabi, from about 2000 B.C. Not only does the Code imply "a high medical organization," according to *Encyclopaedia Britannica*, but the Code identifies hundreds of drugs that were in use during Hammurabi's reign as King of Babylon. One of the drugs was mustard, although there is no indication of how such a drug might have been prepared or administered.

Mustard was so greatly esteemed by the physicians of ancient Greece that they believed it had been discovered by Asclepius. This son of Apollo, having learned the art of healing from the great and wise centaur Chiron, became a celebrated physician, who was eventually worshiped as a god, both in Greece and in Rome.

The Greek philosopher and naturalist Theophrastus may have given mustard its original name, *sinapis*, but it was the Romans who introduced it throughout Europe. Wherever their armies marched, the mustard plant, probably both black and white, accompanied them, the young leaves to be eaten as a pungent, nourishing cooked vegetable or as a fresh salad. The Romans crushed or ground the seeds and steeped them in must, the juice of fermenting grapes, and added this sauce as a seasoning to their meats and fish. The conquered Gauls and Saxons were quick to adopt the plant for this last purpose, and their respective names, *moutarde* in French and *mustard* in English, evolved naturally from the Latin *mustum* and *ardens*, for "burning must."

In the course of the ensuing centuries, mustard became increasingly popular. By the sixteenth century, it was cultivated in English gardens and commonly used as a condiment. Shakespeare alludes to it, both in *A Midsummer Night's Dream* and in *Henry IV*. In 1623, John Gerard said that "the seede of Mustard pounded with vinegar is an excellent sauce, good to be eaten with any grosse meates, either fish or flesh, because it doth help digestion, warmeth the stomache and provoketh appetite."

A few decades later, another English herbalist described what was clearly a thriving cottage industry. "In Glostershire about Teuxbury," he writes, "they grind Mustard seed and make it up into balls which are brought to London and other remote places as being the best that the world affords." This last assertion may or may not have been entirely accurate, because by then, France had become the major source of table mustard—certainly for sale in its own "remote" capital, Paris. Dijon mustards, seasoned with such delicacies as capers, anchovies, walnut paste, mushrooms, or tarragon, won their renown early and have kept it into the present.

All mustard in Europe was formerly prepared as balls, the seeds lightly crushed and mixed with some vinegar or honey and a little cinnamon; the Italians mixed orange and lemon peel with the seeds. The prepared mustard balls were then simply stored until needed. When that time came, the ball was "relented" with a little more vinegar, ready for use. In England, John Evelyn particularly recommended using the "best Tewksbury" or the "soundest and weightiest Yorkshire seeds." Mustard continued to be sold in balls until the eighteenth century.

It was then that an astute, imaginative Yorkshirewoman, Mrs. Clements, of Durham, is credited with inventing the system of preparing mustard powder. It was an invention for which English mustard eventually won world renown, largely thanks to the East India Company. In Mrs. Clements's time, and long afterward, it was known as Durham Mustard. When needed, and only as much as was needed, the powder was mixed into a paste with cold water or vinegar.

In time, as the popularity of mustard grew, numerous methods evolved whereby mustard could be made more or less spicy. By then, the milder white mustard had been introduced to England, and was often mixed with, or substituted for, the black. When both mustards were used together, it was called double mustard. Milder flavors depended on how much flour was added to the powder, and on whether the powder was mixed with vinegar, cold water, or milk. Some examples of the so-called London Mustard contained mostly flour, turmeric for color, and pepper for flavor. And when the "John" Company's ships returned from East India filled with exotic goods of trade, their crews also came back with a taste for curries—and Indian, or brown, mustard, *B. juncea.*

One point, however, was probably agreed on by all—that only cold water (or other cold liquid) should be used for making prepared mustard; heat diminishes or destroys the flavor. In her *Culinary Herbs and Condiments*, Mrs. Grieve explains how the pungency of mustard is released:

> All varieties of mustard seeds contain fixed oil, proteins, and mucilage. From the point of view of their utility as condiments, however, the most important constituents are certain sugar-compounds or glucosides and a mixture of active substances or enzymes called "myrosin." The sugar-compound present in White Mustard seed is different from that in Black Mustard seed. In the presence of water, the myrosin in White Mustard seed acts on its peculiar sugar-compound, producing, amongst other substances, a sulphur-compound of very pungent taste. This substance is practically non-volatile, hence is devoid of aroma.
>
> In the case of Black Mustard seed, treatment with water induces some-

what similar chemical changes, a different sulphur-compound, however, being produced. This substance has not only a very pungent taste, but it is volatile and has a very powerful odour. It is the chief constituent of the volatile oil of mustard.

This oil is, physiologically speaking, very powerful. Its odour is quite unbearable, and, applied to the skin, the oil rapidly produces blisters. Nevertheless, when present only in traces, its odour becomes agreeably aromatic and its flavour is equally pleasant.

In modern-day practice, the black seeds are still used in prepared mustards. The black or the brown seeds are also fried until they pop in the preparation of certain Indian foods, such as *gobhi musallam*, the boiled cauliflower dish, or *vindaloo* beef curry; even for a hot version of the famous East Indian soup known as mulligatawny. Black and brown seeds are both used in French prepared mustards; German and English varieties often combine the black or brown seeds with white mustard seeds (actually, these are nearer yellow). By far the best-known use of the white seeds, of course, particularly in America, is in the bright yellow mustard that traditionally accompanies hotdogs. The white seeds are also most frequently added to pickles, mixed spices, and chutneys.

Among the Christmas gifts I often make is a selection of mustards. Depending on the tastes of the recipient, they are mild, spicy, or downright roof-of-the-mouth-burning hot. For a milder taste, I use the white seeds, because they lose some of their pungency with time. The kind of liquid used also makes a difference—wine or vinegar for a strong flavor; water or fruit juice for a milder taste. Another way I reduce the heat is to mix cornstarch or arrowroot with the mustard powder; adding a little sugar or honey mellows the flavor, and turmeric heightens the color.

To make a fairly mild mustard, I finely grind 5 tablespoons of white seeds. In a small bowl, I mix them until fully blended with the grated rind (without the pith) of half each lemon and orange, 5 tablespoons cold water, 2 tablespoons each cider vinegar and lime juice, plus ½ teaspoon sea salt. I cover the bowl and let it stand at room temperature overnight. If the mustard is too dry, I adjust its consistency by adding a few drops of cider vinegar or water, depending on how mild I want the mustard to be. Then I transfer it to small, screw-topped glass jars, and share it with a deserving friend. Of course, if I wanted to make this mustard really tame, I could add the water hot.

For a mustard that will bring tears to your eyes, on the other hand, I finely grind 3 ounces yellow mustard seeds and 1 ounce black. In a small bowl, I thoroughly blend the ground seeds with 1 scant tablespoon ground ginger, 2 cloves finely minced garlic, 1 tablespoon each dried rosemary, thyme, and oregano, ½ teaspoon sea salt, 1 tablespoon Worcestershire sauce, 3 tablespoons olive oil, ½ cup red burgundy wine, and ½ cup cider vinegar. Again, I leave the mixture covered overnight, then store it in screw-top glass jars, after making any necessary adjustments to the consistency.

In comparison with the fun it is to make one's own prepared mustard, eating mustard greens, cooked or fresh, might be a letdown, if it were not for the fact that they are so extraordinarily healthful. Its greens are an excellent source of vitamins A, B_1, B_2, and C; its flower buds are rich in protein. Besides, they are so available, *free*, throughout the spring countryside! For those who are unable to collect the free bounty, mustard greens are also sold in supermarkets. Their taste is somewhat bitter and peppery, enough so to make them interesting. Only the basal leaves should be gathered, as soon as they begin to appear in early spring. As these and the already bitter upper leaves mature, they become unpalatably bitter.

Washed, finely chopped, and used sparingly, they add zest to tossed green salads. As a cooked vegetable, they should be steamed, or gently boiled in plenty of water, for at least 15 to 20 minutes. However, their bulk shrinks in the cooking process. Mustard greens can be seasoned with a sprig of fresh tarragon or oregano, or served with a white onion sauce seasoned with either of these herbs. The greens can also be served alone, with pepper, salt, and a dab of butter, or as a warm salad with vinaigrette.

Before long, as the flower stalks shoot up, clusters of pale yellowish green buds appear at their tops, very like miniature broccoli heads. These are considered one of the great spring delicacies offered by Nature, and can be gathered from the moment the buds appear until nearly all in a cluster have opened. Barely blanched in a little salted water, or steamed for no more than 3 minutes, they can be served at once, either seasoned with pepper, salt, and a pinch of nutmeg, or with a fine cheddar and white-wine sauce. Not only are they rich in protein and vitamin A, but their flavor, though mustardy, is mellower than that of the leaves.

Even the tender green seedpods are edible. Flavorful yet still quite mild, their special piquancy is relished in mixed salads, omelets, fish stews, and steamed vegetables. The pods can also be pickled, either alone or with beans, onions, or beets.

The most practical method for gathering the seeds is to cut the seed-bearing stalks into a paper bag and then spread the pods on a large sheet in a warm, dry location until they are fully ripened. Either split open the pods by hand, to spill the beans, as it were, or else thresh them with a flail, a spade, or a baseball bat. Even a dustpan will do, if that is all you have. The point of this exercise is to winnow out the seeds. Sift these through a fine wire mesh or a colander, and—presto!—a private hoard of mustard.

The following achievements may not be among mustard's best known, but they have proved invaluable through the ages. They are the weed's ability to sweeten soil, because of the salts that mustard absorbs into itself; its ability to protect collards and brussels sprouts against aphids and flea beetles; and its release into the soil of a magical chemical that suppresses root rot as well as the emergence of destructive nematodes.

Although no part of the mustard plant is considered poisonous, it can become irritant if abused. So it is not surprising that, given its powerful characteristics, mustard has always been employed as a healing plant. John Parkinson observed more than 300 years ago that it is "of good use, being fresh for Epilepticke persons . . . if it be applyed both inwardly and outwardly." Not much later, Evelyn stated that mustard is of "incomparable effect to quicken and revive the Spirits, strengthening the Memory and expelling Heaviness. . . ."

Culpeper considered mustard good for snake poison, if taken in time, and for "drawing out splinters of bones, and other things of the flesh." He also wrote that "it purges the brain by sneezing . . . and stirs up lust" (not necessarily at the same time, of course!). "It is an excellent sauce for clarifying the blood, and for weak stomachs," he claimed, and recommended it for such divers ills as coughs and lethargy, "pains in the sides, and gnawings of the bowels," as well as toothache, gout, sciatica, hair loss, freckles, and the "lousy evil."

Perhaps the most time-honored application is of the mustard plaster. Although it appears to be no longer in use by the medical profession, it continues to find favor in homeopathic practice. For centuries plasters were applied externally in the treatment of chest congestion caused by respiratory diseases or colds, as well as for the relief of neuralgia, sore muscles, stiff, aching backs, sprains, arthritic pains of the joints—even chilblains. Mustard plasters stimulate blood circulation, warm the skin, and help to ease the pain of the affected part.

In fact, the physician who took care of three generations of my family—we always referred to him as "old Dr. Winder"—recommended mustard plasters for a cousin of mine—we always called him "poor Bruce"—who suffered terribly from asthma as a child. I might not have remembered the plasters were it not that old Dr. Winder also told poor Bruce to sleep under an open window, through all weather, in all seasons (I once saw him thinly covered with snow, fast asleep!). Yet, to this day nobody can be sure whether it was the plasters, the open window, or time that cured Bruce of asthma.

The ancients knew never to use mustard plasters on tender, sensitive areas. To prevent possible irritation of the skin, they also never applied the mustard directly to the skin. The preparation my aunt made for Bruce is considered typical even today. She mixed equal parts of powdered mustard and rye flour into a thick paste with tepid water. She smeared the paste on a piece of muslin and covered the paste with another piece of muslin. She left this plaster on poor Bruce's chest only long enough for the skin to begin to redden. Immediately, she removed the plaster, washed the skin with lukewarm water to remove all traces of mustard, dusted the area with a little cornstarch, and covered Bruce's chest with soft, dry cotton clothing.

Footbaths made by pouring hot water over crushed black mustard seeds can not only work wonders for tired, aching feet but can also chase away headaches and colds and heal chilblains. And when the body is chilled "to the bone" from too much time spent outside in rain and cold, or from an impending infection, I believe nothing quite equals the soothing, warming effects—right down to the cockles of the heart—of a mustard bath.

How could I forget either one of these treatments—the plasters and the baths? When we were children, poor Bruce may have had the plasters, but, being subject to colds and chilblains, I got the baths.

Taken internally, ½ teaspoon crushed seeds in warm water, mustard is recommended as a mild laxative and to restore normal function. If being punched in the back, counting backward, or drinking water from an upside-down position have no effect on hiccups, some of my doughty relatives think nothing of drinking a cooled infusion made from boiling water poured over a teaspoon of mustard powder.

Knowing all this, if it has given us no pleasure until now, perhaps we will see this weed through new eyes next spring. And perhaps we will more nearly appreciate the biblical parable, when Jesus said, "The kingdom of heaven is like to a grain of mustard seed, . . . which indeed is the least of all seeds: but when it is grown, it is the greatest among herbs."

Capsella bursa-pastoris

Capsella bursa-pastoris
(S H E P H E R D ' S   P U R S E)

Hardy Annual

COMMON NAMES	Shepherd's Purse, Shovelweed, Lady's Purse, Witches' Pouches, Pickpocket, Mother's Heart, Case Weed, St. James' Wort, Rattle Pouches, Pepper-and-Salt, Clappedepouch, Pickpurse, Poor-Man's Permecety, Toywort, Shepherd's Sprout
USES	Culinary, medicinal
PARTS USED	Whole herb
HEIGHT	6 inches to 2 feet
FLOWER	Small, white, inconspicuous, self-fertilized; March to December
LEAVES	Basal rosette of deeply cut or lobed leaves; arrow-shaped, sessile stem leaves
ROOT TYPE	Taproot
HABITAT	Open waste ground, sandy, poor soils, lawns, fields, garden beds; worldwide, except tropics
PROPAGATION	Seeds contained in numerous heart-shaped pods
CONTROL	Frequent weeding, mowing

Although shepherd's purse grows through most of the year, it makes its presence most apparent in early spring. The first glimpse of this plant every spring is like a flashback to childhood days, to tales of castles and magic spells, to gallant knights and fairy-tale princesses. Suddenly one day, seemingly overnight, it appears everywhere, in garden beds and lawns, in fields and meadows and waste ground, like bands of miniature maypoles adorned with tiny white flowers that turn into heart-shaped "lanterns" dancing on every breeze.

Of course, its fairy-tale magic notwithstanding, it's a weed, and a persistent weed at that—but one that deserves our indulgence as a saver of lives. Those "lanterns"—the seedpods—are the plant's most distinguishing feature, their shape reminiscent of the leather food pouches worn by shepherds in times past. From this comes the name shepherd's purse, not only in English, but also in German (*Hirtentäschel*) and French (*bourse-à-pasteur*). In fact, the weed's botanic name can be translated as "the little box [is] the purse of the shepherd."

In a less tranquil association, the Irish name *clappedepouch* alludes to the former custom of lepers begging at major crossroads. These were among the few public places where such outcasts of society were able to attract the attention and compassion of passersby with a bell or clapper. They received their alms in a wooden cup that was attached to a long pole, reminiscent of a shepherd's purse seedpod at the end of its stalk.

Shepherd's purse is a member of that large tribe, the mustard family. Like many other plants that originated in the fertile regions of Europe and the Near East, it accompanied human travels and migrations around the globe, settling wherever the soil was tilled. Not surprisingly, therefore, it was unknown in the New World before the arrival of the Pilgrims, a fact that was noted by the seventeenth-century English traveler John Josselyn.

Fully grown, this weed rarely exceeds 18 inches in height, although, in response to particularly rich soil, it will stretch to 2 feet. The average shepherd's purse, however, is more likely to be about 9 to 10 inches high. Like the dandelion and chicory, the stem of shepherd's purse rises from the crown of the taproot in the center of a basal rosette of deeply cut or lobed leaves. The few stem leaves are arrow shaped and stalkless, and above them, on the sometimes branched spikes, appear the numerous self-fertilized tiny white flowers.

It's when the flower heads turn to seed that the plant is particularly welcome in the world of childhood fantasies. I remember that my younger sister, Helen, and I used to "invite" various beetles to sit on our make-believe furniture of sedum leaves, and offered them dinners of grasses, peas, and currants under the "lights" of shepherd's purse spikes.

The flattened pods contain a portion of the roughly 40,000 seeds a single plant is capable of producing. Once the pods have opened, these seeds are scattered far and wide by every passing breeze. If this sounds positively frightening, lucky for all of us, Nature's sense of balance sees to it that most of the seeds serve as taste treats for birds and small animals. All the other tiny seeds settle down wherever they fall—and they do this generally not too far from the parent plant— in a garden, a lawn, a pasture, or a cultivated field. Here, their numbers are usually further reduced, through hoeing, plowing, and mowing, once the weed's new crop of rosettes appears in late summer or early spring.

It is the *Capsella* seeds that have recently given us a glimpse into the plant's evolution, showing us what is literally the skeleton in its closet. Scientists have learned that sometime in prehistory shepherd's purse probably confronted a crisis of life and death. Who knows, perhaps the primeval forests of its origins were denuded by the indiscriminate rapacity of dinosaurs, flying reptiles, and/or sundry other herbivores. Perhaps this gradually impoverished the soil to a degree that deprived the weed of even the most basic nutrients. Although the exact circumstances may not be apparent, one thing is evolutionarily certain: This modest weed became what might be termed a closet carnivore, a meat-eating plant.

However, unlike such well-known carnivorous plants as the Venus's-flytrap and the pitcher plant, shepherd's purse indulges in this practice secretly. In order to survive, young *Capsella* seeds under cover of the soil snare whatever protein-rich prey is unfortunate enough to come their way. The method is another of Nature's ingenious inventions. A sticky, sweet substance covering the seeds attracts various microscopic animals that, once trapped, are unable to extricate themselves from their predicament. By absorbing the protein and other nutrients from these unhappy creatures, the *Capsella* seeds gain the necessary strength to develop into vigorous new shepherd's purses.

But, having *taken* life, the weed now possesses the wherewithal to complete the cycle by *saving* life. Although it has never numbered among major healing herbs, shepherd's purse has been considered invaluable for its hemostatic properties since ancient times. However, centuries passed before this ability to stanch bleeding became widely known, largely through the writings of Pierandrea Mattioli, an Italian lawyer and physician who published several herbals as well as a commentary on the writings of Dioscorides, in the 1600s.

The secret of *Capsella's* blood-clotting ability is its content of vitamin K, considered to be an important factor both in preventing and arresting abnormal bleeding. Modern herbalist medicine continues to use the weed for stopping hemorrhages of every kind, be they external or internal, especially of the kidneys, stomach, and lungs. Most often recommended is *Capsella* tea, taken in doses of 1 wineglassful 4 times daily. An infusion is made by adding 2 ounces of the herb to 24 ounces of water, boiling this down to 1 pint, and straining.

As an effective remedy for excessive menstruation, European herbalists recommend drinking *Capsella* tea in doses of 2 cups per day, beginning 8 to 10 days before the onset of a period. They also prescribe this tea to normalize irregularities in the menstrual cycle during puberty. *Capsella* tea is not only used to stem bleeding but it is also recommended for regulating high and low blood pressure.

For an almost instant arrest of nosebleed, many people simply soak a cotton swab with the freshly expressed juice of shepherd's purse and insert it into the affected nostril.

Perhaps the most dramatic usage of shepherd's purse came during World War I. Because other drugs were commercially unavailable to them, German physicians relied instead on this weed for its styptic properties. So it's not unreasonable to say that many a soldier may have owed his survival, at least partially, to a number of protein-rich, microscopic primitive life forms.

Among its other uses, an infusion of shepherd's purse is regarded as an important remedy in catarrhal conditions of the bladder and ureters. In fact, because of its reputed stimulant, diuretic, and antiscorbutic action, the weed has been much used in the treatment of numerous kidney complaints. Many people take an infusion as a refreshing spring tonic, in the belief that it relieves such circulatory disturbances as hypertension, varicose veins, arteriosclerosis, and hemorrhoids, and that its effect is most beneficial in cases of mucous inflammations of the respiratory, digestive, and urinary tracts.

Nevertheless, valuable as *Capsella* tea may be in treating assorted ailments, its taste is less than appealing. Still, wrote a Dr. Ellingwood, who is not otherwise identified, "a little Spirits of Juniper much disguises the flavour. A small quantity of Nitrate of Potash [potassium nitrate] will further disguise it, and not detract from its medicinal value."

Ancient Greeks and Romans found the seeds beneficial as a mild laxative, and the plant was a staple in early English kitchens as an astringent for cases of diarrhea. In treating arthritic pains, poultices made of the plant are said to have had positive results. The leaves, bruised and macerated in cider vinegar and applied as a poultice, have also cooled inflamed swellings.

External wounds washed with a decoction of the weed, it has been claimed, heal within a matter of hours. According to Culpeper, "if bound to the wrists, or the soles of the feet, it [shepherd's purse] helps the jaundice. The herb made into poultices, helps inflammation and St. Anthony's fire. The juice dropped into ears, heals the pains, noise and matterings thereof. A good ointment may be made of it for all wounds, especially wounds in the head."

European herbalists have found that a sitz bath infused with shepherd's purse is particularly soothing for hemorrhoid sufferers and is easy to prepare. Maria Treben suggests adding a strong decoction of the leaves to warm bathwater, which should be deep enough to cover the kidneys as well. At the end of 20 minutes, instead of drying oneself, she suggests getting wrapped up in a terry robe or bath sheet and, covered with a light blanket, sweating for an hour in bed. Whether using shepherd's purse or any of several other herbs, such a bath has long been a common remedy in my family for all sorts of ailments.

Shepherd's purse also plays an important role in a mixture recommended for bed-wetting. Herbalist Richard Lucas suggests steeping for a few hours ½ ounce each of shepherd's purse, agrimony (*Agrimonia eupatoria*), lady's slipper (*Cypripedium pubescens*), corn silk (*Stigmata maidis*), oak bark (*Quercus robur*), and crushed licorice root (*Glycyrrhiza glabra*) in 2 pints boiling water. He concludes, "strain; add two ounces glycerin. Dose: One teaspoonful before each meal and at bedtime."

Capsella, like all members of the mustard tribe, is edible, although I personally do not like its acrid taste. Nevertheless, the plant has been known as a potherb since the days of Dioscorides and Pliny, and probably long before them. The *very* young basal leaves were eaten raw in salads or mixed with such other spring greens as chicory and dandelion in much the same way as modern foragers do. Some modern gatherers also steam or boil the leaves for 3 to 5 minutes, seasoning them with a little butter, a cheese sauce, or oil and vinegar. According to at least one connoisseur of wild plants, Horace Kephart, shepherd's purse is "delicious when blanched and served as a salad. Tastes somewhat like cabbage, but is much more delicate."

Lastly, the seeds, when dried and ground, provide a piquant seasoning known since ancient times and used by many people in the present. Ask yourself, however, when was the last time your food consumption made you a Good Samaritan? Just think, should you wish to do so, your sprinkling the powdered *Capsella* seeds into soups or stews, on eggs or fish or meat, hot or cold, could save a life beneath the surface of the soil!

Chelidonium majus

Chelidonium majus
(GREATER CELANDINE)

Perennial

COMMON NAMES	Celandine, Greater Celandine, Swallowwort, Garden Celandine, Common Celandine, Wartweed, Tetterwort, Felonwort, Grecian May, Killwort, Sightwort
USES	Medicinal
PARTS USED	Whole herb
HEIGHT	To 3 feet; slender, round, downy, branching stems, swollen at nodes
FLOWER	Bright yellow, in small, loose, stalked clusters at ends of stems; blooms all summer
LEAVES	Yellowish green above, pale bluish green underside, paired, deeply cleft, lobed; large terminal leaflet
ROOT TYPE	Branched, fleshy
HABITAT	Hedgerows, edge of woods, waste places, rubble, rock walls, damp, rich, untilled soils
PROPAGATION	Blackish seeds in long narrow pods
CONTROL	Weeding
CONTRAINDICATIONS	Unsafe: Internal use in large doses is poisonous.

helidon is the Greek word for "swallow," and celandine probably derives its name from the fact that it begins to burgeon when the swallows arrive in spring and dies back when they leave again in autumn. Legend has it that swallows use a sprig of this weed, or its juice, to restore the sight of their young when these cannot see. John Gerard debunks this belief, however, based on the writings of Cornelius Celsus. He quotes the famous Dutch botanist as saying that "when the sight of the eies of divers yong birds is put forth by some outward means, it will after a time be restored of it selfe, and soonest of all the sight of the Swallow." More than 1,000 years earlier, Aristotle had made a similar observation.

So much for *that* particular legend, but there are others. Carrying the weed on one's person, together with the heart of a mole, is supposed to enable the wearer to vanquish his or her enemies and also to win lawsuits. Another legend, citing Pierandrea Mattioli, holds that "worn in their shoes that have the Yellow Jaundice so as their bare feet tread thereon it [celandine] will help them of it." However, what lends the weed a downright air of perversity is an old belief in southern Europe that celandine sings when a sick man is about to die and weeps when he is going to live. Just how the plant weeps and sings is not explained.

Celandine rarely grows too far away from human habitations. It is particularly fond of locations that simultaneously offer partial sun and shelter from the worst assaults of wind and weather. It favors the sides of buildings, for instance, a hedgerow, the edge of woods, the base of a fence, or the crevices in a rock wall. Even so, sometimes this weed is found growing off the beaten track, in a forest clearing or amid scattered rubble. If found there it is of particular interest to archeologists. In Europe, at least, more than once, celandine's presence has been a decisive clue to the discoveries of ancient settlements.

Once established, the weed is not easy to remove. This is especially true if it has lodged itself in a rock wall, where its roots become firmly anchored. Because all parts of the herb are extremely tender and brittle, celandine does not readily concede defeat or relinquish its hold. This is true even when the herb is seized in the firmest grip of the most practiced gardener, who almost invariably ends up holding no more than the broken leaves in his or her hand. For this reason, unless the soil is extremely friable, I rout celandines of all sizes with a digging fork. It avoids stress—mine.

Nevertheless, because of its bushy growth and its rich, golden flowers, *Chelidonium majus* is, in fact, a rather attractive plant. I keep it beside the garage, where it provides me with one of the earliest signs of spring and with a lingering touch of green at the end of autumn. Gerard likened the celandine's leaves to those of columbines, only more tender and jagged, describing the flowers as being "in shape like those of the Wal-floure." These are succeeded by long, narrow seedpods in upright clusters. "The whole plant is of a strong unpleasant smell," he added only too accurately, "and yeeldeth a thicke juice of a milky substance, of the colour of Saffron."

Celandine was greatly valued by old herbalists, who employed it for numerous ailments. Whether it was to be used fresh, or dried for winter use, they recommended harvesting it between May and July. Modern herbalists suggest the root be gathered in spring and fall, the herb throughout the summer.

Gerard claimed that "the juice of the herbe is good to sharpen the sight, for it cleanseth and consumeth away slimie things that cleave about the ball of the eye, and hinder the sight, and especially being boiled with hony in a brasen vessell, as Dioscorides teacheth." With something akin to mystical fervor, Culpeper asserted, "This is an herb of the Sun . . . one of the best cures for the eyes, for all that know any thing in astrology know that the eyes are subject to the luminaries; let it then be gathered when the Sun is in Leo, and the Moon in Aries . . . let Leo arise, then may you make it into an oil or ointment." Not for a moment did he seem to doubt "that most desperate sore eyes have been cured by this only medicine."

Indeed, there appears to be some validity to these claims. Until a few decades ago, and possibly even today, cotton pads dipped in a lukewarm, weak infusion of celandine and applied to the eyelids were used by English country people to soothe sore or puffy eyes. In Germany, many herbalists treat cataracts by placing a fresh, washed leaf of celandine directly on the closed eyelid overnight. In fact, one of the German common names for the plant is *Augenkraut*, or eye herb.

Drinking an infusion of celandine was formerly recommended for its diuretic, purgative, and tonic effects, as well as for its reputedly beneficial effects on jaundice and various skin disorders. For obstructions of the liver and gall, and to help in cases of edema and scabies, the herb is said to have been decocted in wine, with a few aniseeds added to the brew.

However, it is difficult to imagine anyone willingly drinking the potion— even when it is disguised with aniseed. Not only is the taste of celandine juice unpleasant, it is also extremely acrid. Moreover, it can cause severe inflammation of the digestive tract, as well as irritation of the mucous membranes and of the nervous system.

Not surprisingly, therefore, modern herbalists recommend taking greatest care with internal doses of celandine. Or, in referring to its internal use, as Mrs. Quelch phrased it more bluntly, "it had better be left alone, at least as a home medicine."

Nevertheless, celandine remains a favored remedy in homeopathic preparations. Under these controlled conditions, it is prescribed for such varied problems as gout, jaundice, liver disorders, and catarrhal and arthritic complaints. Celandine has also proved to be an active herbal remedy of inefficient glandular function, poor circulation, and obesity. Not least, it is regarded as one of the finest preventives of the formation of gallstones.

Used externally, the plant's orange-yellow juice has been respected for centuries, especially when applied undiluted to the edges of small open wounds, which it is said to heal rapidly. But to counteract the possible irritation of surrounding healthy skin tissue, Culpeper recommended dabbing this with vinegar. European herbalists of today apply soothing calendula ointment. To prepare the ointment: In a saucepan over medium heat, add a handful of coarsely cut calendula flowers, leaves, and stems to 4 ounces melted lard, or other setting fat, stirring constantly until the mixture foams; remove from heat, cover tightly, and let cool overnight. On the following day, lightly warm the pan to soften the fat, then press through muslin, and store the ointment in tightly closed jars. To make a lotion, substitute almond, olive, or sunflower oil for the lard.

On the other hand, celandine's irritating effect on the skin has a long and respected tradition in the removal of warts and corns. These are "burned" off in a few days by daily applying a few drops of the juice directly on the affected area.

For the purpose, the juice is usually expressed from fresh leaves, flowers, or stalks that have first been rinsed in water. According to German folklore, however, this treatment is wholly successful only if the juice is applied to wart or corn under the pensive gaze of a waning moon.

Perhaps the most consistent use of celandine in folk medicine has been its application to a variety of skin problems, ranging from eczema and ringworm to pimples, scabs, and blisters. It is from these disorders, which were formerly known as tetters, that the weed earned one of its common names. German herbal practice treats tetters with applications of a wash (1 teaspoon herb steeped 30 seconds in 1 cup boiling water, strained, and cooled) or of the juice mixed with an equal amount of water.

A soft paste made of the juice mixed with powder of brimstone was formerly used to remove discolorations of the skin. Boiled in lard, the roots or the leaves and flowers of celandine are an old English country remedy for hemorrhoids. In Russia, celandine used to be—and perhaps still is—a popular medicine for the treatment of cancer, used both internally and externally. *Geschwulstkraut*, yet another common name, which may be translated as "swelling herb" or "tumor herb," suggests that celandine may have been put to similar use in Germany.

To ease the pain of a toothache, a weak infusion or decoction of the herb swilled through the teeth is a time-honored folk remedy. Chewing the root against toothache was recommended by Gerard. In some rural areas of England earlier this century, a kind of combination of both these treatments was employed, by applying to the painful tooth a wad of cotton soaked in a celandine infusion. And in modern English herbal cosmetic usage, powdered dried celandine leaves mixed with a little honey and charcoal made from burned toast are considered an effective tooth-cleansing powder.

In the days of Queen Elizabeth I, placing the powder of the dried root on a hollow or loose tooth was an accepted method of removing the tooth, or causing it to fall out. In fact—it may be no more than hearsay, of course—it is said that the monarch herself once avoided the agony and indignity of an old-fashioned tooth extraction with pincers, by applying celandine to one of her decaying "black pearls," as her dentition was euphemistically (not to mention sycophantically) described; and shortly afterward she removed the offending jewel with her own fingers.

But even if you choose not to put celandine to any of its uses, perhaps next spring you will see a swallow dipping gracefully toward a weedy patch, and, an instant later, soaring away with a sprig of green in its beak. If you do, who knows, you may give life to an age-old legend.

Chenopodium album

Chenopodium album

(L A M B ' S - Q U A R T E R S)

VITAL STATISTICS

Annual

COMMON NAMES	Lamb's-Quarters, Goosefoot, Pigweed, Wild Spinach, Fat Hen, Allgood, Frost Blite, Mutton Tops, Dirtweed, Baconweed, Dirty Dick, Muck Hill, Midden Myles, Dung Weed, Melgs
USES	Culinary, medicinal
PARTS USED	Young shoots, leaves
HEIGHT	1 to 6 feet, sometimes taller, erect, slender, much-branched stalk; often mealy white-green
FLOWER	Small, green, in densely clustered terminal spikes; sometimes turn reddish; tiny, brownish black, glossy seeds; June to October
LEAVES	1 to 4 inches long, wedge shaped, somewhat lobed or toothed, dull grayish green on top, mealy white undersides
ROOT TYPE	Long, white, taproot, usually branched
HABITAT	Rich, moist soils; waste ground, fields, manure and compost heaps; around barnyards
PROPAGATION	Seeds
CONTROL	Frequent cultivation, weeding

There have been times when I have thought I would go mad if I heard about yet another wild plant that "tastes just like spinach." I have no objection to spinach—far from it—but too much of anything can become boring. Besides, quite often my taste buds have disagreed with the analogy. Naturally, therefore, I was skeptical when a friend in the country was about to introduce me to an unfamiliar vegetable at dinner one day. She called it pigweed. "It tastes like—" she started to say. "Spinach?" I finished. "You know it already?" she responded, sounding disappointed. I shrugged. "Just a guess." The very name, *pigweed*, did not inspire confidence or excitement in me. But I tasted it, mostly to be polite.

Shortly afterward, I was greedily asking for more. This "taste-alike" wildling was even better than spinach, its flavor more full-bodied. "Pigs are nobody's fool," I commented, and secretly decided that no pig living anywhere near me would henceforth get even a snort at this weed. Of course, it was an idle threat—I still resided in New York City at the time.

Nevertheless, in my zeal I transplanted several field-grown pigweeds to my city vegetable patch, where they died almost at once. At the time I didn't know that they prefer to start from seed and rarely survive transplanting. I also tried to interest my city friends in spending a pigweed–picking day in the country, hoping they would see the humor of the "theme" and later thank me for enriching their lives. But the name put them off. It was *their* loss, I reflected; they lacked the spirit to discover new worlds. So I went on alone, eating pigweed at every opportunity. Still . . . if only it had a more appealing name.

It had, as I soon learned—*lamb's-quarters*. Although there is no logical basis for it, nor an indication of who first used it, or why, *lamb's-quarters* sounds decidedly more charming than does *pigweed*. *Lamb's-quarters* also distinguishes the plant from any other pigweed—an important point, considering that *pigweed* is a name applied to numerous different plants in different regions around the United States.

All sixty species of the genus *Chenopodium* are called *pigweeds*! They are also known as *goosefoots*, supposedly because their leaves resemble the webbed feet of geese, *Chenopodium* being the Latinized version of the Greek for *chen* (goose) and *pous* (foot). To add a touch of confusion, this genus is merely one of 100 genera in the huge goosefoot family, the Chenopodiaceae.

This clan originated in the regions around the eastern Mediterranean. In the course of history, many of its members migrated to the temperate zone throughout much of the world, including North America. It is probably fair to say that most of them are considered weeds; a modest number are annual or perennial garden ornamentals, and several species are not only edible but represent food staples or favorite foods. Among this latter group are at least three "serious" vegetables—beets, Swiss chard (a kind of beet), and—spinach!

With connections like these, it is not surprising to learn that lamb's-quarters has been in use as a food throughout its history, albeit not always for humans. What *is* surprising is the fact that, despite its abundant growth everywhere, its worth is generally so little known. Instead, it is altogether ignored, if not, at best, sometimes confused with one or another of its numerous relatives.

Of these, perhaps the most respected is a blood brother, so to speak. Good King Henry is the name, *C. bonus-henricus*, although this plant is far less prevalent and not really considered a true weed. Nor is it named after King Henry VIII, as is often stated (in fact, it is unclear how the word *king* slipped into the name at all), but after a virtuous goblin in Germanic folklore, "good Henry," *der gute Heinrich*, who helps maids with their housework in exchange for a humble saucer of milk. (It is also unclear which plant, if any, may be named after this goblin's malicious folkloric twin.) Good King Henry, one of the perennial goosefoots, has long been cultivated; it is available at some nurseries. Its height rarely exceeds 1 foot, with a somewhat creeping growth habit, and its dark green leaves are lance shaped. In my admittedly biased opinion, the flavor of *this* goosefoot cannot approach that of lamb's-quarters.

Without question, lamb's-quarters can be a serious nuisance, especially if it is allowed to produce seeds. It is able to grow in the poorest ground and to withstand prolonged droughts, if it must; it is also one of the last weeds to be killed by autumn frosts. In brief, it is a weed with an extraordinary capacity for survival. This was demonstrated by seeds recovered from an archeological site, where they had been deeply buried for some 1,700 years. Incredibly, like Sleeping Beauty, they awakened and began to germinate as soon as they were touched by Nature's "kiss" of water, soil, and air.

Yet, given the chance, lamb's-quarters always prefers rich, moist, cultivated earth. For this reason, it is also one of the most reliable indicators of good soil. Helped by what must be its own form of radar, lamb's-quarters always finds its way into potato fields. In moderate numbers, it is also considered to be a useful companion plant to corn. Of course, it has no compunction whatever about invading home garden beds; however, being highly visible there, you can quickly rout it through regular weeding. Better still, plow or dig it under before it blooms: It will add valuable fertilizer to your soil.

More than anything, lamb's-quarters likes the rich, lazy, luxuriant lifestyle it can enjoy in compost and manure heaps. Here it can thrive effortlessly, drawing on all the moisture and nutrients deep in the pile, all of them made so generously available by you and/or a nearby barnyard. In the process, alas, it tends to halt the decomposition of the heaps, so it is best to let the weeds wilt before adding them to the compost.

Its lust for the nutrients in the soil or in the compost or manure pile is not entirely selfish. Because of its long taproot and its predilection for cultivated soil, lamb's-quarters can reach to greater depths than many of its companion plants, and thus release nutrients from the lower regions into the upper strata. Other plants benefit, the soil benefits, and we benefit, especially if we eat lamb's-quarters. Its nutritive value is considered not only equal to but even greater than that of spinach: Besides its large content of iron, calcium, and albumen, lamb's-quarters is also rich in vitamins A and C.

Lamb's-quarters used in poultry feed helps to produce both dark-yolked eggs and tender meat. In fact, numerous American chickens of my acquaintance have been known to flutter, squawking, over high wire fences in order to charge a patch of this weed, fresh or seedy. Canadian farmers used to cultivate the plant for their sheep and pigs. Farmers in India raised the plant for themselves—the greens for fresh vegetables and the seeds for meal. Although the weed is not cultivated in Europe, nor in the United States—enough of it grows wild—nature lovers and foragers count it among their favorite foods. American Indians, particularly in the Southwest, have always known it as both a food and a medicine.

It is considered good for preventing iron deficiency, boosting energy, and being a useful source of vitamin C. It is cited to possess mildly laxative and diuretic qualities, and to act as an antiscorbutic. In my own experience, the leaves, boiled, cooled, and applied to sores, cleanse these and promote their healing. The American Indians applied poultices of the leaves to reduce swellings.

Mostly, however, lamb's-quarters is used for eating. We collect only the tender young plants, before they exceed 10 inches, or the tips and young leaves. By doing so, the plants serve double duty: They are prevented from producing seed that spreads around the garden while providing what my family, numerous friends, and I look upon as one of the most delicious and versatile vegetables—*free!* If the soil is reasonably good, there is little danger that lamb's-quarters will ever completely forsake it. The seeds and their chief carrier, the wind, will see to that.

One of the major advantages of using lamb's-quarters is that it does not require double cooking, as do so many other weeds. Even when the greens are thoroughly rinsed they do not absorb the water; it simply rolls off them. Added to a little boiling, salted water, covered, and cooked for 7 to 10 minutes (or steamed for the same length of time), lamb's-quarters has long been a popular vegetable among the American Indians of Arizona and New Mexico, eaten either alone or with other vegetables. Many people also enjoy eating lamb's-quarters leaf tips raw in salads. Raw, the young leaves are often added to sandwiches in lieu of bean sprouts, lettuce, or watercress; chopped, they go well with chicken or egg salads, or as garnish with roasts, cold cuts, or cheese platters.

To obtain the optimal benefit of this weed's full-bodied, somehow nutty flavor, I simply steam the greens, lightly season them with pepper, salt, and/or a pinch of nutmeg, plus the juice of half a lemon. Some people prefer vinegar. Still others add butter, or serve the greens with a cream or cheese sauce.

Chopped lamb's-quarters can be added to stews, prepared like nettle soup (see page 241), substituted for spinach when making a soufflé, or baked in "spinach" pie. They can be used to stuff savory mushrooms or veal. Puff pastries can be filled with chopped lamb's-quarters and shrimp in a wine sauce, or they can be rolled into an omelet or savory crepe. The possibilities are boundless.

Not least, I personally like to have a supply of lamb's-quarters for winter as well. For this reason, I blanch and freeze several packages of meal-sized portions. With the help of these, February and March somehow pass more quickly.

Among birds, lamb's-quarters seeds are an ever-popular food; among humans, the following is more limited, although the seeds are highly nutritious and surprisingly versatile. The American Indians have always known this. They use them as a cereal and cook them into a breakfast gruel. They also dry the seeds and grind them into a flour, which resembles buckwheat in color and in taste, with which they make bread.

Lamb's-quarters seeds are smaller than mustard seeds, but they form in such profusion that gathering them is one of the easiest harvesting tasks of all. In late autumn, when the seed clusters are dry, I hold a paper bag under them and simply strip the seeds into the bag. Depending on the number of available plants, I can gather large quantities of seed within a very short time. Afterward, I rub the dried husks between my palms and sift the seeds.

Whole and raw, or lightly crushed, the seeds can be used for hors d'oeuvres, for example, by mixing a small quantity into cream cheese or any other soft cheese, as one would do with coarsely ground pepper. Whole and lightly roasted, I sprinkle them on cheese omelets together with chopped chives. Or I sprinkle them on the top of a bread loaf or rolls, before placing these in the oven.

The seeds ground into meal can be mixed with white or whole wheat flour, to be made into pancakes and waffles, muffins and rolls, herbal bread and corn bread.

There are times now when I wonder how on earth I managed to survive before I met lamb's-quarters. Today, the weed is so normal a part of my diet that, during the winter months, I sometimes find myself absentmindedly wondering whether the local supermarket has "run out of" it, because I cannot find it in the produce department. Last year, I planted one row of lamb's-quarters in the vegetable garden; this year, perhaps it should be two. On the other hand, if my supplies of the weed run low, I can always buy spinach. After all, it really does taste like lamb's-quarters.

Chrysanthemum leucanthemum

Chrysanthemum leucanthemum

(OXEYE DAISY)

VITAL STATISTICS

Biennial

COMMON NAMES	Daisy, Oxeye Daisy, Dog Daisy, Goldens, Marguerite, Field Daisy, Great Oxeye, Dun Daisy, Horse Daisy, Bull Daisy, Butter Daisy, Maudlin Daisy, Moon Daisy, Horse Gowan, Whiteweed, Gowan, Maudlinwort, Rising Sun, Dog Blow, Dutch Morgan, Fair-Maids-of-France, Poverty Weed, Moon Penny, Margaret, White Golds, Poorland Daisy
USES	Culinary, medicinal, cosmetic
PARTS USED	Whole herb
HEIGHT	1 to 2 feet, stems erect, mostly unbranched
FLOWER	Single, white, about 2 inches across; central disk yellow, depressed in middle; May to August
LEAVES	Dark green, narrow, smooth; coarsely, irregularly lobed; stalked leaves in basal rosette, stalkless along stem
ROOT TYPE	Perennial, whitish, fibrous, slightly creeping
HABITAT	Fields, roadsides, waste places, sunny locations
PROPAGATION	Seeds
CONTROL	Cultivation, soil improvement

Few sights are cheerier than a huge assemblage of daisies in an unused pasture, or clumped beside a country road, faces raised to the sun, and swaying to the rhythm of a warm and gentle breeze in June. It is no wonder that nineteenth-century poet William Cox Bennett saw in them "a smile of God."

Alas, not everybody feels so enthusiastic about this weed. Yes, it *is* considered a weed—a beautiful one, but still a weed—in spite of its relationship to that large and divers clan, the chrysanthemums. It is found throughout Europe, North America, and parts of Asia, predominantly as a wild plant, but often welcomed into garden beds.

The great botanist Linnaeus seemed to be determined to leave no room for confusion when he named it *Chrysanthemum leucanthemum*. The names are derived from three Greek words: *chrysos*, meaning "gold," *anthemon*, meaning "flower," and *leuc*, meaning "white." Together they give almost equal emphasis to the flower's two instantly recognizable features, the ring of snow-white petals around the wide central disk of golden yellow. Clumsy as "white-flowered gold flower" may sound in translation, there is no doubt about which plant it identifies. Far less adamant are the weed's common names, except the word *daisy* itself. Derived from the Anglo-Saxon *daeges-eage*, it means "day's eye," and identifies the plant as precisely as any modern scientist could wish.

However, throughout the daisy's history, popular opinion about it has clearly never been unanimous. In some areas it has been called by the sinister name of devil's daisy; elsewhere it was once believed to be among the sacred plants of Saint John the Baptist. The ancients of Greece dedicated it to Artemis, the Olympian goddess of Wild Nature and also of women, while the name *dun daisy*, by which the plant is still sometimes known in southwestern England, is a lingering reminder of its connection with the Thunder God in Anglo-Saxon tradition. The name *moon daisy* again alludes to the goddess Artemis. In northern regions, the plant was known as Balder's brow, for the god of light and peace in Teutonic mythology, whereas Christians named it maudelyn or maudlin daisy, in honor of Saint Mary Magdalen.

Marguerite is the name by which the weed is known throughout most of Europe and in Russia. The English sometimes still also call it fair-maids-of-France. Both these names can be traced to the French Margaret of Anjou. For her marriage to England's King Henry VI, in 1445, the fifteen-year-old princess chose a simple spray of three oxeye daisies as the motif to be embroidered on her wedding robes.

In far less lofty circumstances, many of us, as children, have undoubtedly asked the daisy to lead us to a prince or princess of our own. I know I did, as did my childhood friends. Earnestly, we each picked a daisy blossom and then set about plucking its petals, one by one. At the same time, we recited—just as earnestly—the magic incantation, "he loves me, he loves me not," no doubt sounding like junior Druids. When I think of all the daisies I denuded! And they never got it right anyway. Whoever "he" was at the time didn't even notice me.

All in all, I think it is safe to assert that the daisy is generally more often loved than hated. Since its introduction by early settlers, the oxeye daisy has spread throughout the United States and is probably scorned only by dairy farmers whose pastures it has invaded. However, the farmers' reasons are perfectly valid: Cows refuse to eat the plant; daisies in the hay mean, in effect, less hay for the cows; that, in turn, means less milk production and, consequently, less income. On the other hand, North Carolinians thought so well of the daisy that, until 1941, they designated it the official state flower.

There is no denying that great invasive hordes of daisies can cause problems. For example, their presence could ruin a beautiful lawn. But to the lawnkeeper who is alert to Mother Nature's ways, the presence of daisies is also an indicator that the soil is beginning to turn sour and in need of some attention. Such a lawnkeeper knows that repeatedly raking the lawn to loosen the surface soil and matted grass and roots is a useful first step, and may even be enough to correct the situation. If not, one or more applications of lime and bone meal on the thoroughly raked lawn will solve the problem.

In regularly cultivated, tilled, or plowed ground, where the soil is rich in neutral compost and humus, where the plant's roots are frequently churned up, and the seeds prevented from developing, the daisy soon disappears.

Of course, like me, you may rout daisies from the lawn, but actively introduce them in the perennial border. There is no doubt that clumps of these wide-eyed blossoms somehow unify an entire bed, linking the various colors and textures, giving the whole a distinctly individual air of charm and grace.

However, I suggest keeping their numbers in check, or they will spread at an alarming rate. Near the end of every summer, therefore, I usually pull up at least half the leaf rosettes that have developed from last year's seeds. These rosettes are easily distinguished by their larger size from this year's newly emerging crops. If the leaf clusters are in the wrong place, I use them to replace old clumps, which, in any case, tend to develop a weedy, tatty appearance. Such regular weeding and grooming really does make them look as "fresh as a daisy" should look, while still providing ample supplies for bouquets.

Because of the visual pleasure the daisy always gives me, be it outdoors or in a vase, I tend to forget that there is also a serious side to this weed. For example, its young and tender basal leaves add an interesting, uncommon flavor to salads when used sparingly. Theirs is probably an acquired taste, blending bitterness with a balsamic quality, which may be the reason why daisy leaves have failed to enjoy the same popularity of such other spring greens as dandelions and sorrel.

That same blended flavor, I suspect, also gives daisy wine its special personality. This wine has been traditionally made from the blossoms, exactly like dandelion wine (see page 223). The source of the bitter taste is an acrid juice that permeates the entire plant, including the flower heads, where it is found in the ring of green sheaths under each bloom. Not only do these sheaths support the blooms but, perhaps more important, they prevent insects from gaining access to the nectar from below.

Far more committed is the medicinal side of the daisy. As an herbal remedy, it has been used since ancient times. Its action is regarded as both tonic and diuretic, as well as antispasmodic, and its effect is described as similar to that of camomile. Although it has never gained the renown of other weeds used in folk medicine, the daisy continues to be included in modern homeopathic treatments for various afflictions.

Culpeper described the daisy as "a wound herb of good respect, often used in those drinks and salves that are for wounds, either inward or outward." He also recommended applying the bruised leaves to reduce swellings. He firmly believed it very fitting for the daisy "to be kept both in oils, ointments, and plasters, as also in syrup." He advocated the use of daisy to refresh the liver, to temper the heat of choler, and to cure ulcers of the mouth. Not least, "the juice of the daisy," he wrote, "dropped into the running eyes of any, doth much help them."

For yet another remedy, Culpeper suggested preparing a decoction made of daisies, "of wall-wort and agrimony, and places fomented or bathed therewith warm, giveth great ease to them that are troubled with palsy, sciatica, or the gout." The same, he claimed, also healed "bruises and hurts that come of falls and blows." And "an ointment made thereof doth wonderfully help all wounds that have inflammations about them," such as "those for the most part that happen to joints of the arms and legs."

According to Gerard's writings, Dioscorides already said "that the floures of Oxeie made up in a seare cloth doe asswage and washe away cold hard swellings, and it is reported that if they be drunke by and by after bathing, they make them in a short time well-coloured that have been troubled with the yellow jaundice." A similar belief was carried over in some rural areas of England, where, until the late nineteenth century, people still drank a decoction of the fresh herb boiled in ale to cure jaundice.

The entire plant has been traditionally used, including the roots, although these found favor only in the United States, in earlier days. An extract was prepared from the roots for the treatment of tuberculosis. Most often, however, it is the leaves, flowers, and stems that are employed, fresh or dried. Herbalists generally agree that the best time to harvest these is in May and June, and that the flowers should be gathered only when they are freshly opened, either to be used at once or to be air dried and stored.

A simple tea made with the flowers (1 teaspoon fresh or ½ teaspoon dried to 1 cup boiling water) is said not only to relieve chronic coughs and bronchial catarrhs, but also to soothe whooping cough. Especially among European herbalists, sniffing such an infusion is often recommended for nasal congestion or the discomfort of a head cold, as is the inhalation of an herbal steam made with daisy flowers.

A similar infusion of only the flowers or of the entire herb is often recommended as a wash and compress for skin eruptions and irritations, as well as for badly healing wounds. Because of the plant's acrid juice, applying an infusion to the skin is an effective insect repellent. I find it useful also simply to rub the fresh herb directly on my skin.

The daisy can even help us obtain our own "fresh as a daisy" look. To get rid of spots and pimples, the freshly expressed juice from the stems is applied at bedtime and washed off the following morning. Or a simple wash—prepared by steeping 1 handful daisy flowers in 1 cup boiling water, kept covered until cool—can be splashed on face and neck as often as desired, and should be left to air dry. According to folk belief, not only does this treatment keep the skin smooth, it also prevents the blotching brought on by age.

To soothe and smooth roughened skin—be it on the hands, face, elbows, legs, arms, or knees—some herbalists suggest steeping a handful of freshly picked flowers in warmed milk for 5 minutes, then applying the strained herbal milk to the affected area. This is said to be particularly beneficial when applied after a shower or bath at bedtime and rinsed off the next morning.

The flower heads can also be dried, ground into powder, and used to prepare a soothing lotion. Four ounces warmed milk are poured over 2 teaspoons powdered daisy. Stirred into the mixture until dissolved are 1 tablespoon each of borax, bicarbonate of soda, and glycerine. Decanted into a tightly capped and labeled bottle, the lotion can be refrigerated. This simple preparation is considered particularly beneficial for skin that has been roughened by prolonged exposure to harsh winds, weather, and/or water.

According to country traditions, even tired eyes and swollen eyelids can benefit from a soothing compress or wash made of daisy flowers prepared in a strong infusion. If such a lotion is bottled and cooled in the refrigerator, it can be used on a compress for the almost instant relief of eyes that have been wearied and strained by too much light and sun after a day at the beach, in the mountains, or working outdoors.

And yet, its serious side and earnest endeavors aside—even the fact that it acts like a true weed at times—I personally believe that the daisy has no need to prove itself. It *is*, and God smiles. That's enough for me.

Cichorium intybus

Cichorium intybus
(CHICORY)

Perennial

COMMON NAMES	Chicory, Succory, Witloof, Wild Endive, *Barbe-de-Capucin*, Blue Dandelion, Hendibeh, Blue Sailors
USES	Culinary, medicinal, commercial
PARTS USED	Whole plant
HEIGHT	2 to 5 feet
FLOWER	Blue, sometimes pink or white; similar to dandelion; June to October
LEAVES	To 12 inches long, jagged, grooved, hairy, bluish green; emerge directly from root, forming rosette on soil surface
ROOT TYPE	Taproot
HABITAT	Roadsides, waste places, wild gardens; chalky soil, full sun
PROPAGATION	Seeds, roots
CONTROL	Regular weeding

hicory has been known and respected since prescientific times, and cultivated for at least 5,000 years for its medicinal and culinary properties, which are still valid today. The name *chicory*, or *Cichorium*, is believed to be of Egyptian origin, and it was known as *chicourey* to ancient Arabian physicians, perhaps already in the time of the great code maker Hammurabi, some 4,000 years ago. With only minor modifications, the word *chicory* has traveled through the millennia and been absorbed into virtually every European language. The plant's specific name derives from *hendibeh*, another Eastern word for chicory.

Slightly corrupted, *hendibeh* has reached us as *endive*, and—instantly, I sense a flood of unvoiced questions rushing toward the produce section of your local supermarket. "You mean I spend top dollar for a *weed?*" I hear you sputter. Seems so. You see, the loosely bunched and bitter salad greens available at supermarkets are one kind of endive, pronounced like "hive." However, equally endive—only much more chic and expensive, and always packed like fine bone china—are the blanched and crisp, tightly leaf-wrapped heads shaped like 5 inch shuttles, which are pronounced in the French fashion, "ahndeeve," even though it is the Belgians who raise them as a major cash crop by the name of *witloof*. In any case, both types are the forced leaves of certain types of chicory, and, no matter which you prefer to eat, you're in good company.

Such historic gastronomes and herbal practitioners as Pliny, Ovid, Vergil, and Horace all mention chicory in their writings as both a vegetable and a salad ingredient. Well before these venerable Romans came the Greek naturalist Theophrastus, who wrote that the plant had been in use among the ancients. And that was from *his* perspective—Theophrastus died about 287 B.C. Nearly 1,000 years later, Charlemagne demanded that chicory be one of the seventy-five herbs planted in his gardens. And a few centuries later, the appeal of chicory remained unabated—Elizabeth I drank a broth made from it, although the plant was not much cultivated in England in her day. In the New World, President Thomas Jefferson considered chicory "a tolerable sallad for the table," while in France the use of chicory root for making a coffeelike brew was known and loved long before coffee itself was even heard of there.

In fact, it is quite possible that when coffee was first introduced into Europe in the sixteenth and seventeenth centuries, the French at once recognized the kinship in flavor between chicory and the real thing. And so it's perhaps not too farfetched to suppose that they stretched their scarce (and undoubtedly expensive) supplies of the new commodity with additions of the familiar and readily available chicory. Certainly, this mixture remains a typically French custom even now, as chicory gives coffee a darker color and a slightly bitter taste. This is not to say that all Frenchmen drink this brew; *au contraire*, a Dr. Leclerc, it seems, contends that chicory "transforms the most delicious mocha into a bitter pharmaceutic potion that makes a gourmet's taste buds stand on end."

There seems to be little dispute, however, as regards the value of chicory drunk to combat a variety of ailments, or as a depurative, a digestive, and a restorative. The seventeenth-century English herbalist Nicholas Culpeper recommended chicory for numerous afflictions, including "for swooning and passions of the heart." John Parkinson, the director of the Royal Gardens at Hampton Court in the same century, called it a "fine, cleansing, jovial plant." Who knows, had he been born a century earlier, in the time of Henry VIII, Parkinson might have advised his sovereign to sip a cup of succory for yet another of its many vaunted benefits—the relief of digestive disorders!

In the United States, that insatiable scholar and naturalist Thomas Jefferson was already raising chicory during the American Revolution, from seeds he is believed to have imported from Italy. In 1795, he extolled the plant's value as cattle fodder to George Washington, calling it "one of the greatest acquisitions a farmer can have." And still, although especially sheep would have agreed with Jefferson (they are particularly fond of chicory), it wasn't until the massive European migrations to these shores during the nineteenth century that chicory came to settle here in such numbers as to be regarded by many as an indigenous weed.

Today, the conspicuous blue of chicory blossoms is so common a sight along country highways and byways from June to September, even into October, that we take the plant quite for granted and would never imagine the ways in which we might introduce it into our everyday lives. Nevertheless, the cultivation of chicory has been throughout its history an often major economic factor, and remains one even today in parts of Europe—and the United States.

Perhaps most immediately endearing is chicory's habit of rarely invading our flower beds or *any* regularly cultivated area. Although chicory multiplies freely, it does not usually transgress beyond its accustomed place, where it thrives, neither unobtrusive nor intrusive, seemingly quite at peace with its fortunes. If it chooses to live on your property at all, chicory will seek out a dry and sunny location to establish its colonies of leafy rosettes. Rising from these basal leaves are stiff, hairy stems and branches. The lower leaves are large, raggedly lobed, oblong, and hairy, their size rapidly diminishing toward the upper portion of the stems, where they cling like stumpy blades of grass to the axils of the stems. Consequently, the stalks, all of them at haphazard lengths, present a naked and gawky appearance, reprieved only by the flower heads clustered in the axils.

Its brilliant blue flowers are chicory's greatest charm. You might find them lighting up a steep embankment or encircling an old barn—dancing starbursts briefly captured. Most often, however, you will find the flowers spilled and scattered beside a country road or among fields, suspended randomly along the plant's skeletal branches like blue crystals worn in memory of a richer, happier past.

In form not unlike the dandelion, chicory flowers bloom for only a few hours each day. So dependable did he discover them to be of opening at five o'clock in the morning and closing five hours later, at his latitude, that Sweden's renowned botanist Carl von Linné, better known as Linnaeus to horticulturists everywhere, included chicory in his floral clock at Uppsala. In England, chicory blossoms generally open one or two hours later than in Uppsala, and close about noon. Here in my corner of the Northeast, outside my study window, I can observe the chicory flowers open between seven and eight o'clock in the morning and begin to close, their color fading rapidly, from noon onward.

Left in its wild state, chicory favors sandy, gravelly, well-drained soil, which makes removing it difficult, except after a heavy rain, because of the plant's long, divided taproot. Thin, woody, and brittle, this brownish root is somewhat less palatable than is its fleshier cultivated sibling, although it lends itself every bit as well to making a beverage. Paradoxically, however, should you decide to cultivate the chicory for any of its several culinary and medicinal uses described below, you will be positively astounded at how eagerly it responds to compost-enriched, friable soil. Compared with extracting them in the wild, pulling chicory roots out of such a medium is no more arduous than harvesting parsnips from the vegetable garden—and like parsnips is exactly how young chicory roots are cooked and eaten in Belgium and elsewhere in Europe, as well as in this country.

All parts of the chicory plant can be used, but by far the most widely known use is of chicory roots as an adulterant of, or substitute for, coffee. In some sections of Europe and the United States, coffee simply isn't coffee without an admixture of chicory. To obtain this admixture, the long taproots are dug up in late spring and early summer (before they become too woody). First they are thoroughly scrubbed and then roasted slowly in the oven at a setting not above 200 degrees Fahrenheit. The roots are usually cut crosswise, into roughly 1-inch lengths, to accelerate the drying process. When the pieces are dark brown and brittle, they are stored in a tightly closed container, ground finely or coarsely only when needed for the freshest flavor. I prepare the chicory brew exactly as I prepare coffee, but use it more sparingly, because chicory has a somewhat stronger, more bitter flavor. If it is to be mixed with coffee, a practice that the French believe serves as counterstimulant to the excitation caused by coffee, then it is advisable to experiment with the proportions to suit the individual palate.

I remember the special treat it was for us children when we were allowed to join the grown-ups with our cup of chicory "coffee" after dinner. And another vague recollection that defies my placing it in a particular location or time evokes the rather pleasant taste of chicory boiled in milk—a sort of noncafé au lait.

For those who decide to give chicory "parsnips" a try, it might be wise to consider doing one of three things: (a) get some help in preparing them; (b) dine alone; (c) first cultivate the chicory roots in your garden. The reason for this is very simply that, by the time the tough rind has been peeled from a wild chicory root, there is not much of it left to cut crosswise or to cook. Although such first-year roots are, in fact, very tasty, the amount of work involved in getting them from the soil to the table somehow doesn't seem worth it all. (Unless, of course, it's the only food prospect you've had in over a month.) By contrast, the chicory roots cultivated in the vegetable garden swell to respectable garden-sized vegetables.

If the young leaves are gathered from practically the first moment they appear, there is nothing quite like chicory, a most welcome and nutritious spring green. Best known by far are the tender white leaf crowns that emerge subsurface atop the roots. These crowns, when dug in spring, make a superb salad or steamed vegetable. Thoroughly rinsed, they can be dressed as a salad with a little vinaigrette or served as a vegetable with melted butter, pepper, and salt. These same blanched leaf crowns have long been the chief ingredient of a popular wild and bitter winter salad in France called *barbe-de-capucin* (beard of the Capuchin monk).

The blanched leaf crowns can be easily cultivated wherever chicory is plentiful by covering the young plants with upended flowerpots whose holes have been blocked against all light. Before long, blanched chicory leaves begin to show under the covers. Or, chicory roots can be dug up, trimmed of leaves and stems, and bedded in a box of moist sand in a basement, a root cellar, or any other location where frost cannot reach them. This method yields one to two crops of blanched crowns during the winter.

For an early-spring fresh green vegetable, I particularly enjoy the slightly bitter but refreshing taste of the very young aboveground chicory tops, especially when they are steamed or boiled, and served either with a vinaigrette or with a hollandaise sauce. Alas, chicory greens turn bitter all too quickly, once they are exposed to light.

Chicory flowers are rarely mentioned for culinary use, although they are used to add both color and a stronger flavor to their blanched sibling leaves in a salad. But I have encountered at least one recipe that briefly gives these blue blossoms a center-stage place in the kitchen. The fact that the flowers fade so rapidly once they are picked is, no doubt, a major reason for the absence of culinary uses. Nevertheless, the recipe (or *receipt*, as it would have been called then) that I include below is from a little-known work by a Sir Hugh Plat, entitled *A Closet For Ladies*, which appeared in 1608. "To make Conserve of Cichory flowers," he writes:

> Take of your Cicory [sic] flowers new gathered; for if you let them [stand] but an hour or two at the most, they will lose their colour, and do you very little service; therefore weight [weigh] them presently, and to every ounce of flowers you must take three ounces of double refined Sugar, and beat

them together in a Mortar of Alabaster and a wooden Pestle, till such time as they bee thoroughly beaten; for, the better the Flowers and Sugar be beaten, the better will your conserve be; let this always be for a generall rule; and being well brayed [pounded?], you must take them up, and put it into a Chaser [?] cleane scoured, and set it on the fire till it be thoroughly hot: then take it off, and put it up, and keepe it all the yeare.

With the addition of violets, this same conserve was known as "violet plate" in the time of King Charles II, and was used in the treatment of tuberculosis (known as consumption in those days).

In Switzerland, exactly seventy years after Sir Plat's publication, Italian lawyer and physician Pierandrea Mattioli offered his recipe for a similar confection. In this instance, the flowers were first finely chopped before being mixed with the sugar and pounded in a mortar; instead of being cooked, Mattioli's preparation was merely transferred to a porcelain pot and placed in the sun until the sugar was liquefied, then stored. This confection, wrote Mattioli, "strengthens the heart . . . opens, cleans, and strengthens the liver, banishes heartburn, stops fevers and incipient dropsy, cools the inflamed liver and all internal organs. In brief, this sugar serves all infirmities."

Medicinally, chicory has been most often associated with benefiting hepatic problems—"the friend of the liver" is what the Greek physician Galen called the plant 2,000 years ago. More modern herbalists, from Culpeper on, seem to agree that a cup of the prepared decoction of chicory root (1 ounce boiled in 1 quart of water for 5 minutes, then left to infuse for another 10 to 15 minutes) drunk before meals, three times daily, rids the body of all impurities. Chicory has been credited with sharpening the appetite, stimulating digestion, relieving urinary infections, and aiding in the excretion of bile. Herbalists have also long considered the decoction valuable for diabetics, because it lowers the blood-sugar level, and for sufferers from arthritis, gout, and edema.

In England, syrup of succory has long been used to act as a nonirritant laxative for children, in doses of 1 teaspoon each up to three times daily, depending on the child's age. The syrup is prepared by extracting and straining the juice from fresh, crushed chicory roots, then gently simmering equal parts of juice and sugar (or honey) until a syrupy consistency is attained. For brief periods, such a syrup can be stored in the refrigerator, in a tightly closed container, but is best made fresh as needed.

Thirty grams of the entire plant boiled in 1 liter of water for 5 minutes, then strained and pressed through cheesecloth, is what my French great-grandmother Aurélie is said to have used for treating jaundice. In addition to administering this in doses of 3 wineglasses per day for 3 days, she apparently also insisted that the patient must simultaneously chew a fresh sage leaf for optimal results.

Not least, the freshly bruised leaves are used as a poultice to ease swellings and inflammations, pimples, and sores. The crushed leaves are especially beneficial, according to Culpeper, if a little vinegar is added to them. Culpeper also recommended an infusion of chicory leaves to treat inflamed eyes.

I realize that all the foregoing makes chicory almost too virtuous for our skeptical modern minds. And yet, surely, there must be truth in much of it; how else would chicory have been able to keep its ancient reputation intact, in spite of the temporary oblivion to which it has been sent during the past century? Even so, the most cursory study of modern homeopathic practices quickly reveals that ranked *highest* among the thirty-eight flower remedies espoused by Dr. Edward Bach's revolutionary treatment methods of human ailments is chicory, *C. intybus.*

Daucus carota

Daucus carota
(Q U E E N   A N N E ' S   L A C E)

V I T A L   S T A T I S T I C S

Biennial

COMMON NAMES	Queen Anne's Lace, Wild Carrot, Bird's Nest, Bee's Nest, Devil's Plague, Crow's Nest
USES	Culinary, medicinal, cosmetic, commercial
PARTS USED	Whole plant
HEIGHT	2 to 3 feet, stems upright, branched, furrowed, tough, hairy
FLOWER	Tiny, white; densely clustered flattened heads atop separate flower-bearing stalks; *solitary tiny red or deep purple flower at center;* June to October
LEAVES	Finely divided, alternate, grayish green
ROOT TYPE	Long, slender taproot, lateral rootlets, creamy colored, fleshy, smells of carrot
HABITAT	Fields, roadsides, meadows, pastures
PROPAGATION	Seeds
CONTROL	Digging up root, repeated mowing, plowing, or tilling

A field of Queen Anne's lace in bloom is a beautiful sight in summer, disks of white flowers swaying gently on tall, slim stalks, like doilies threaded among the grasses of waysides and meadows, a rich prospect for bouquets and dried flowers. But to the farmer, who may think he barely blinked since the last time he plowed that field, such a sight is a headache, an invasion by one of the weediest of all weeds.

First of all, the farmer most likely calls it wild carrot. Because of its deep taproot, its presence tells him that the soil in this field may be well worth planting with good crops. He already knows too well that if he does nothing, a single wild carrot plant may develop and broadcast several thousand seeds in the course of one season. And he knows that once it is firmly entrenched, not only will wild carrot crowd out his haying grasses but he will have only one sensible method of regaining control: He must plow the weed under, before it can bloom, so that it turns into compost, which will nourish the soil. And if he keeps planting sweet clover or other crops, eventually he will succeed in banning the weed from this particular field. Of course, he also knows that other wild carrots will be waiting on the sidelines for a chance to return to center field.

Wild carrot, or Queen Anne's lace, is a member of the Umbellifera or Parsley family, which contains more than 1,500 species, including such familiar edibles as fennel, carrots, parsnips, celery, parsley, dill, chervil, and caraway. It also contains several poisonous plants—the poison hemlock (*Conium maculatum*), water hemlock or spotted cowbane (*Cicuta maculata*), and fool's or dog's parsley (*Aethusa cynapium*).

Although all of these plants, the good and the bad, bear at least a family resemblance to one another in their growth form, leaves, and flowers, the wild carrot is, without question, the most easily recognized, by the dark red or purple, almost black, dot in the center of its white flower heads. That dot is, in fact, a tiny flower, the plant's most distinguishing feature, and its assurance of reliability.

Nearly 2,000 years ago, Pliny declared that this plant "grows spontaneously; by the Greeks it is known as *Staphylinos.*" But that was already the second of three names the Greeks had for it. Around 500 B.C., the comic poet Epicharmus had referred to it as *Sisaron*. In Pliny's own time, the Athenian Dioscorides accurately described the vegetable we know as carrot, calling it *Elaphoboscum*. Meanwhile, back in Rome, Pliny's compatriots used the single name *pastinaca* to denote both the carrot and the parsnip. Based on the verb *pastinare*, "to dig up," this name undoubtedly alluded to the long, fleshy roots of both vegetables, and surely created some gastronomic confusion. Not surprisingly, in the second century A.D., the Greek physician Galen must have wanted to put an end to such misunderstandings when he named the carrot *Daucus pastinaca*, to distinguish it from the parsnip, which eventually became *Pastinaca sativa*.

However, the naming game was not yet over. In what seems to be the first specific mention of the garden carrot, the word *carota* appears in the writings of Athenaeus, a Greek scholar of the second century A.D. Thirty years later, the Roman epicure Apicius used the same name in his manual on cookery. Then, although it is unclear why common usage from the sixteenth century onward reverted to *Daucus*, Linnaeus finally put an end to it all (but did he?), by assigning *Daucus carota* as the official botanic name—to both the wild and the garden carrot! And what do modern Greeks do? Why, they call the carrot *karoto* and the parsnip *dauki*. Time out for screaming.

Today, there is probably not a country or region where the carrot is unknown. It is believed to be a native of Afghanistan and neighboring Asian lands, and is distributed throughout the temperate regions of the world, in both its wild and garden forms. Cultivated around the Mediterranean Basin long before the Christian era, carrots were widely known and used in Germany and France, as well as in China, by the thirteenth century.

England came to know the carrot quite late. It was not until the reign of Elizabeth I that it was introduced there by Protestant Flemings who had fled their homeland to escape the religious persecutions of Spain's King Philip II. Inevitably, in England, as it had everywhere else, the carrot soon became a popular vegetable, raised in the gardens and fields throughout the country. Before long, it had made its way into fashionable circles, where its delicate leaves became an adornment for ladies' headdresses.

By the time Queen Anne came to sit on England's throne, a legend was in the making in which the dark red blossom at the center of the flower heads symbolized a drop of the royal blood, supposedly formed when the monarch pricked her finger while making lace. Presto, the humble carrot became forever touched by romance and magic. Naturally, by then, again as everywhere else, most people had also learned that the carrots not eaten in their first year of growth went to seed in the second and escaped to the wild.

Probably before Queen Anne was even born, early colonists had brought carrot seeds with them to the New World. As ever, the carrot wasted little time in populating this continent too. Today, it is almost impossible to imagine an American landscape without Queen Anne's lace, the delicate flower-head clusters of summer, like myriad parasols, slowly furling up into the shape of bird's nests to protect the ripening seeds before these fly on the wind to conquer new territories.

Wild or cultivated, carrots are a rich source of vitamin A, and have been valued for their antiseptic and other medicinal properties since ancient times, although wild carrots are considered medicinally superior. All parts of the herb are used—leaves, roots, and seeds—in the form of tea or poultice, juice, pulp, soup, or vegetable. Taken internally, their action is described as diuretic and stimulant, as well as laxative. They are considered beneficial for liver and kidney disorders, for ulcers and skin disorders, and for coughs and flatulence. Carrots also have a respected tradition as treatments for edema and gout. Historically, they have been recommended for the loss of voice and for ridding the skin of spots and blotches. Not least, they are known to improve night vision.

How mothers and grandmothers everywhere are born knowing all this is a mystery, but I have clear childhood memories of being told by one or the other: "Eat your carrots; they're good for you, they'll give you rosy cheeks!" To me personally, having rosy cheeks wasn't all that important; I ate carrots because I had to. When I became a mother myself, there I was, mindlessly repeating that rosy-cheek admonition, without knowing what truth, if any, there was to it. All that changed the more I learned about weeds, particularly about one of my favorites, Queen Anne's lace, the ancient forebear of the common carrot.

The root of the wild carrot has neither the rich orange color nor the fleshy thickness of the vegetable we buy at the supermarket. Instead, it is creamy white and quite slender, much nearer the garden parsnip to look at and touch, but with a much stronger, more aromatic fragrance than the cultivated carrot. It also differs in taste, being pungent, slightly bitter, and lacking the mellow sweetness of the carrot. And much of the time, it is entirely too tough to eat.

Whether for culinary or medicinal purposes, the weed's first-year roots are considered best. In order to retain most of their valuable elements, wild carrots (and the garden varieties) should never be peeled; rather, it is best to scrub or scrape them. As a food, wild carrot roots are usually best served as a cooked vegetable rather than eaten raw, because of their strong flavor. They are simply cut and prepared like ordinary carrots. Dried and ground, wild carrot roots can also be prepared as a pleasant coffee substitute, in the same manner as chicory (q.v.).

According to folk usage, the itching from skin diseases and the pains of boils, varicose ulcers, abscesses, burns, and scalds can be quickly relieved by directly applying the freshly grated pulp of the roots to the affected part. Old-time herbalists also applied such a poultice to ease the pain of cancerous ulcers.

Few remedies are believed to alleviate flatulence as effectively and pleasantly as a cup of a strong decoction made from the fresh or dried roots. To be drunk hot, such a decoction is suggested by herbalists for a persistent cough and asthma, as well as for stomach ulcers. A strong tea is also recommended for kidney stones, whereas a simple infusion of the root is given as a mild laxative, and to improve night vision.

Medical observations in France have indicated that people who drink a wineglassful of the freshly expressed juice of carrot roots first thing every morning not only develop an immunity to upper respiratory infections but also seem to avoid winter colds, bronchitis, and flu.

The green herb is also credited with possessing healing properties. An infusion of the herb (1 ounce fresh or ½ ounce dried), steeped and covered for 5 to 10 minutes in 1 pint boiling water, has been for centuries considered of value in kidney diseases, bladder ailments, and edema. This tea was also recommended for gout, especially if drunk first thing in the morning and again last thing at night. Applied as a wash, or on a compress, the lukewarm tea is a well-known remedy for taking away the burning pain and itching of chilblains.

Credited with antiseptic properties, the green herb is also used externally, particularly to cleanse and treat open sores and ulcers, by being crushed fresh and

macerated in honey, then applied directly to the afflicted part. The herb is credited with preventing the sores from festering, and the honey acts as a soothing
balm.

Even the seeds of the wild carrot are anxious to prove that their energies and
ambitions are not limited to propagation of their species. Since ancient times, they
have been known for their mildly aromatic fragrance and pleasant taste, as well as
for their service as an effective medicine. Centuries ago, they were regarded as
particularly beneficial in treating such divers ills as chronic coughs, jaundice,
diarrhea, and even hiccups. In many country regions, it was not at all uncommon
to believe in eliminating intestinal worms by slowly chewing ⅓ to 1 teaspoonful
of thoroughly bruised wild carrot seeds in one or more doses. Infused in malt
liquor, the seeds were deemed especially potent in the treatment of serious vitamin
C deficiencies. Of course, many people make a simple infusion of the crushed
seeds in boiling water, for no other reason than their enjoyment of such a tea.

Even the little red flower has been credited with medicinal virtues. Chewing
it, according to an old folk tradition, was believed to prevent epileptic seizures.

However, life is not just about food and medicine; it is also about beauty. And
here, too, the carrot—wild or cultivated—stands ready to do its share in keeping
at bay the spots and blotches and wrinkles caused by time and the sun. To help
clear up acne, many herbalists suggest drinking up to 1 pint of carrot juice per day
or mixing it with spinach—6 ounces carrot juice to 2 ounces blendered spinach.

Although I frequently drink carrot juice, facials number among my special
self-indulgences, and my garden provides me with plenty of choices for skin
treatments. For one of the simplest and most pleasant facials, especially when I
have been in the summer sun too long, I extract about 4 tablespoons mixed juice
from wild carrot roots and parsley. I thicken the juice somewhat with a little
honey and yogurt, and, after thoroughly washing my face, apply the mixture for
twenty to thirty minutes, then rinse with lukewarm water, and pat dry. The effect
is marvelous; my skin tingles, feels more elastic and supple. I feel years younger,
and so does my face.

All in all, I have learned how right they were, my mother and grandmother,
when they insisted that carrots are good for me, that carrots will give me rosy
cheeks: They are, and they do.

Equisetum arvense

Equisetum arvense
(FIELD HORSETAIL)

Annual and Perennial

COMMON NAMES	Field Horsetail, Horsetail, Common Horsetail, Bottlebrush, Scouring Rush, Shavegrass, Pewterwort, Dutch Rushes, Paddock-Pipes, Joint Weed, Shavebrush, Devil's Guts, Bull Pipes, Joint Grass, Corn Horsetail
USES	Medicinal, cosmetic, household
PARTS USED	Green, sterile, whorled stems
HEIGHT	*Fertile Shoots.* 8 inches, single, chubby, flesh colored, short lived; blunt-tipped conelike formation on tip; *Sterile Stems.* 12 to 20 inches, green, erect or decumbent, jointed, grooved, deciduous; branches bushy, brushlike, ascending, coarse, pointed, regularly whorled
FLOWER	None
LEAVES	None specifically (see text)
ROOT TYPE	Slender, dark brown, hairy, creeping, branching
HABITAT	Damp woods, roadsides, fields, stream banks, waste places, embankments, swampy areas
PROPAGATION	Spores and rootstock
CONTROL	Very difficult! In-depth soil improvement, repeated cultivation; removing *all* rootstock
CONTRAINDICATIONS	Should be used sparingly

I magine a world without flowers, a world of vast silent forests, where trees like giant ferns grow more than 100 feet tall, where there are no birds, no rabbits, no cattle or horses, no bears or lions or rats. Imagine a world where the first dinosaur will not evolve yet for nearly 100 million years, where the earliest small mammals will not appear for over 1 million years more than that, and where the sixth day of the story of Genesis, when God created man, is nearly 300 million years in the future. Imagine such a world, and you will have entered the earliest ages of the earth, the world of the horsetails, which grew then in abundance and which have come down to us as the only survivors from the great carbon forests of the Paleozoic Era.

The horsetails, roughly twenty-five species of them around the world, are believed to be the most primitive of the fern families. They are literally in a class by themselves, comprised of only the genus *Equisetum*, from the Latin *equus*, "horse," and *saeta*, "bristle." In their original form of enormous plants, they constituted a large proportion of the early vegetation on earth. What they left behind for us to discover were thick layers of coal in which impressions have been discovered, clearly showing that the horsetails have not changed in shape through the millennia, only in size. Most common of them all today is the field horsetail. Although this constitutes a single species, *E. arvense* (from the Latin for "field" or "land"), it has developed so many variations that there are seventeen different-named forms of it, a fact that gives even experts a moment's pause.

All the horsetails share similar properties, but it is the field horsetail that is usually collected for them—perhaps because it is the species most often met with. In botanic terms, the plant is a virtual Fort Knox of silica, which is a compound of the two major elements of the earth's crust—oxygen and silicon—and which is also the main constituent of the earth's rocks. So great is the content of silica in horsetail that when the green stems and branches are burned in a hot, still flame, there remains a white skeleton of silica showing the original structure. Or, examined through a magnifying glass, the green branches reveal tiny crystals of this same mineral, to which we also owe such valuable stones as opals, amethysts, quartz, and agate, among others.

Its content of silica is also the reason for some of horsetail's other common names. John Gerard already explained this in part, more than 400 years ago, when he described it as "small and naked Shave-grasse, wherewith Fletchers [arrow-smiths] and Combe-makers doe rub and polish their worke." Cabinet makers did the same. In the sixteenth century, the German Hieronymus Bock wrote of horsetail that "the maids need [it to clean] the dishes, especially what is made of pewter or other metal," while 3,000 miles away, the American Indians called the plant they used to scour utensils by a name meaning "it is round." And it was its use by whitesmiths, the craftsmen who finished the metals their colleagues had forged, that gave horsetail the name pewterwort.

Horsetails grow mostly in the temperate zones of the Northern Hemisphere, and they decrease in numbers toward the poles and the equator. (A single tropical species grows to a height of 30 feet or more even today.) In a brilliant scheme of deception, horsetails manage to delude us into believing that they are annuals, because the aboveground parts usually are—they definitely are in the case of field horsetails. What they do not indicate—and what we are too naive to suspect—is the fact that the underground roots and rhizomes are not only perennial and long lived but may be exercising that fact of life as much as *several feet* below the surface!

In brief, the horsetails—all of them—are an extraordinary lesson in survival. In this chapter, however, only *E. arvense* is the center of attention; therefore, following are some of its survival tricks. Short, chubby, *fertile* stems appear for a brief time in early spring. They are flesh toned and noded, and often develop stubby, whorled branches from the tight sheaths of dark, lance-shaped teeth around each node. Emerging like a cone from the top of the fertile stem is the *strobilus*, the plant's storehouse of spores, encased in a snug armor of usually hexagonal scales. A stalk connects the center of each scale to the cone's axis, so that the image is not unlike that of a man (cone) holding a shield (scale) at the end of an extended arm (stalk). When the strobilus has matured, it lengthens slightly, and, as though offering a prehistoric preview of flowers bursting into bloom, it splits its armor and neatly releases each hexagonal piece, thereby exposing the spores these carry. Having done their duty, the fertile stems wither and are almost immediately followed by the *sterile* stems.

As is all the growth of horsetails, these stems are symmetrical. They are upright, cylindrical, and generally hollow, except at the nodes. The regularly whorled branches are merely shorter, thinner, slightly less cylindrical versions of the stems. The outer surfaces of both are grooved, and gritty from the silica particles they contain. Clasped around each node is a scale, or *leaf sheath*, which is sometimes called a leaf. Each leaf sheath is cut into several sharply pointed teeth. The branches emerge from the base of the leaf sheaths, and, in turn, have their own nodes and leaf sheaths. Yet the field horsetail never rebranches, unless it is injured. Its ascending branches are bushy and flat topped, except for the central stem. However, as I have already mentioned, the growth forms of field horsetails are quite variable.

Except for the diabolical depths to which they are prepared to sink— literally!—horsetail roots and rhizomes are structured very much like the above-ground growth. The rhizomes are grooved and ridged and have nodes and internodes, and the roots and rhizomes also sprout in whorls from the bases of the nodes. The rhizomes even contain inner-air cavities similar to those of the above-ground growth. And if horsetails find themselves in particularly inhospitable terrain, the rhizomes are fully prepared to develop tubers as additional sources of nourishment.

Digging up the roots, even assuming that this is possible at great depth, can be a thankless and exasperating task. The rootstock breaks very easily, and, quite frequently, a root that may have traveled solo for perhaps 3 feet suddenly sprouts another root that intersects the first at right angles, like a crossroad—and God alone knows where *this* one is headed.

The harsh reality must be faced that horsetail can be a serious pest under certain circumstances, its venerable life story notwithstanding. It is a curse to the farmer whose fields it has invaded and to the generally rare gardener who watches its determined progress into his garden. The immediate lesson to be learned from the presence of horsetail is that it indicates insufficient drainage. Horsetail loves a wet footing, be this in the form of marshland, wet meadows, ponds, lakes, slow-moving rivers, damp woods, rushing brooks, or streams. Granted, most of these conditions are unlikely to be gardened or farmed. However, wet conditions also include underground springs or a high water table, which *can* pose a problem.

Providing good drainage is an essential start that should be closely followed by some serious, in-depth soil improvement, by means of thorough—possibly repeated—plowing, tilling, or digging, and with the help of generous applications of organic matter, to sweeten the soil and add humus to it. Clearly, another important ingredient in any get-rid-of-horsetail scheme must include the meticulous removal, every spring, of every single fertile spore-bearing stem, the moment it appears.

The simple truth remains: Getting rid of horsetail is not easy. The plant is here to stay, as its history has amply demonstrated. Therefore, we may as well see whether it has any uses, and what these are.

Whether or not they knew of the numerous mineral salts contained in horsetail, or what they were, the Romans regarded the plant as a valuable restorative. Modern herbalists agree. They look upon the plant as a natural remineralizing agent in the realm of health. They recommend it not only for its curative but also for its antibiotic properties, knowing that silica can be directly assimilated by the body and knowing that the plant holds an important place in the traditions of folk medicine.

Horsetail is judged to be particularly beneficial to people suffering from anemia or general debility. Its action is characterized as diuretic and astringent. It is prescribed in the treatment of kidney and bladder disorders, arthritis, gout, and skin afflictions. It is recommended for gastric complaints and inflammations of the respiratory tract. It is said to promote urination and stop bleeding, to reduce fevers, calm an overactive liver, and ease nervous tension. It has been used to clear heavy head colds and to soothe inflamed, swollen eyelids. And throughout history it has been relied on to cleanse and heal wounds. In fact, Dioscorides is quoted as having been convinced that "Horse-taile being stamped and laid to, doth perfectly cure wounds," to which Galen is said to have added, "yea, although the sinues be cut in sunder." In similar vein, even modern studies indicate that fractured bones heal more quickly with the help of horsetail.

Culpeper, too, greatly valued the plant's healing powers. "It is very powerful to stop bleeding either inward or outward," he wrote. "It solders together the tops of green wounds, and cures all ruptures in children." In the nineteenth century, the German herbalist Father Kneipp claimed that the plant performs extraordinary service in the treatment of rotting wounds, even cancerlike tumors, explaining that "it draws out, dissolves, and cauterizes the unhealthy matter." He also unequivocally declared horsetail to be "unique, irreplaceable, and invaluable" in cases of hemorrhaging, bladder complaints, and kidney stones and gravel.

American herbalists regard horsetail not only as a reliable diuretic and an effective treatment in all urinary disorders but also as a nutritive tonic. In Germany today, the plant's remarkable effects have also won recognition from practitioners of conventional medicine, who acknowledge not only its usefulness against bleeding but more especially in healing kidney disorders. Repeatedly, they have found horsetail to be successful in the treatment of wounds, abscesses, eczema, and other skin disorders, whether the plant is used in compresses, poultices, or baths.

Only the plant's aboveground, sterile, green growth is ever used, no matter what service is required of horsetail. These tails are harvested from the time the young shoots harden, around the end of May, until they begin to wither, in October. The green tails are cut with knife or secateurs to avoid gathering pieces of the dark root, which should never be included, just as the fertile shoots are regarded as useless. If the tails are to be dried, they are best spread thinly on brown paper and left to air dry, or else placed in the oven set at 180 degrees Fahrenheit (with the oven door ajar), until they are dried. If the dried herb turns brown or gray, it is considered useless and should be discarded.

For internal use, horsetail is generally recommended as a fairly weak tea, either infused or decocted. The latter method is preferred when using the dried herb, because the silica contained therein is less soluble than in the fresh herb. The recommended ratio for the decoction is 4 to 6 *grams* horsetail brought to a boil and simmered in 1 pint water for 5 to 15 minutes. For an infusion, 8 ounces boiling water over 1 teaspoon finely chopped fresh horsetail is allowed to steep no more than 30 seconds. Regarded as most beneficial of all is the freshly expressed juice of the herb, especially in spring. However, because of the plant's high mineral content, some herbalists believe that excessive internal use of it may irritate the kidneys and intestines of some individuals.

To stem heavy bleeding, traditional German usage suggests taking 1 to 2 cups of the tea at once. Some German herbalists even recommend drinking daily 1 cup of horsetail tea as a cancer preventive. In doses of 1 cupful five or six times daily, for one week only, they regard drinking a horsetail infusion as an excellent treatment for edema, and prescribe it, in daily doses of 1 cup, for arthritic pains as well. And to stop a heavy nosebleed, they consider nothing as efficient as inhaling a horsetail decoction through the nostrils.

The same herbalists also regard herbal baths made with horsetail as invaluable in healing wounds, removing gravel from urine, and easing arthritic inflammations. For this last complaint, herbalist Maria Treben recommends no more than one 20-minute horsetail bath per month (3 ounces herb steeped 12 hours in cold water, and strained; heated, this infusion is added to warm bathwater) to cover the kidneys. On the other hand, such a bath, together with 2 daily cups of horsetail tea, she writes, is beneficial for kidney stones and gravel; similarly, she adds, a daily horsetail bath, together with an overnight poultice placed over the afflicted body part, is recommended for cancer of the kidney.

Moreover, many European herbalists since the time of Father Kneipp, and perhaps since long before him, believe that, for inflammations of the bladder, no remedy can compete with a horsetail steam. A steaming hot decoction (4 teaspoons herb boiled 5 minutes in 1 quart water) is placed under the seat of a wicker chair, upon which one sits unclothed but covered with a towel or blanket, for 10 minutes. Kneipp limited herbal baths to three per week.

For certain ailments, Treben recommends using horsetail internally, in the form of a tea, as well as externally. Such external uses of the herb might take the form of a poultice, a compress, or an herbal bath. Also useful is a simple wash (2 teaspoons herb infused 2 to 3 minutes in 1 pint boiling water) into which a compress can be dipped. For a poultice, she suggests, 2 handfuls of coarsely chopped horsetail in a sieve are held *over* boiling water; when the herb is heated, but not too hot for comfort, it is tipped at once onto a cloth and laid directly on the afflicted area, held in place with a warm dry towel. Such a poultice, writes Treben, can be left on overnight. As an alternative, a compress dipped in horsetail infusion and wrung out can be applied, also covered with a warm, dry towel.

The European tea-and-poultice treatment is considered particularly beneficial for lung complaints, urinary disorders, cancerous and noncancerous tumors, and inflamed joints. In these instances, 1 cup of tea before breakfast and another before the last meal of the day is recommended. The tea alone, taken in two or three daily doses of 1 cup each, is also credited with providing relief from the discomfort of hemorrhoids. Treben recommends the poultice alone, applied overnight, for badly healing wounds, for liver and gallbladder complaints, as well as for stomach and intestinal cancers.

Considered an excellent remedy for inflamed gums is gargling frequently with a horsetail infusion (1 teaspoon herb steeped in 1 cup boiling water for 1 minute, strained, and cooled). Horsetail is even regarded as of value in improving the skin. A strong decoction of the herb applied to the face, and left to dry for 15 to 20 minutes, is said to help control and remove skin blemishes. For a skin-tightening effect, horsetail facials are used to control oily skin and large pores (2 tablespoons herb steeped 15 to 30 minutes in 1 cup boiling water, strained, mixed with 1 teaspoon honey, lemon, or cider vinegar). A horsetail decoction can also be used as a daily face rinse on oily skin.

Fingernails, too, can benefit from horsetail, based on English herbalist tradition, by drinking a cup of weak horsetail tea before bedtime, a practice that is said to strengthen the nails. Even chronically perspiring feet are said to pull themselves together in the presence of horsetail. To accomplish this feat, 8 ounces herb are macerated in 1 quart rubbing alcohol for two or three weeks, the mixture shaken every few days, then strained into a bottle, and applied when needed.

However, this multiple great-grandparent of plantdom offers not only us humans the benefit of its remarkable properties but some of its own kind as well. Being itself extremely resistant to fungus growth, horsetail is capable of curing mildew and other fungus infections in a number of plants. Many roses, fruit trees, vegetables, and grapevines have become as familiar with horsetail tea as have many humans—except that they receive theirs in the form of a spray. The spray is both gentle and fast acting, without damaging the plants or the soil. For an effective treatment-spray to combat the leaf curl that afflicts peach trees, however, horsetail is combined with stinging nettle. Both of these biological sprays are prepared by boiling 2 ounces herb (equal parts, in the case of horsetail and stinging nettle) in 1 gallon water for 20 to 30 minutes, and cooling before use.

After so much goodness from this plant, how on earth can we, in effect, bite the hand . . . ? True, there is no question that, if for no other reason than its survival, the horsetail fully deserves our reverence and wonder. True, too, that nobody, surely, who has read anything at all about horsetail would skimp on giving it the respect it deserves. But realistically—if we've dried enough horsetail for a lifetime supply of teas, baths, gargles, poultices, compresses, and lotions, if we have scoured all our pots and pans, rubbed all our wooden furniture to a high gloss, sprayed fungus off the entire neighborhood's roses, closed the large pores on our faces, and stopped our feet from sweating—if horsetail keeps on invading our fields or garden, we have no choice but to call it and treat it as a weed.

Galium aparine

Galium aparine

(CLEAVERS)

Annual

COMMON NAMES	Cleavers, Clivers, Goose Grass, Bedstraw, Catchweed, Hayruff, Scratch-Grass, Gripgrass, Robin-Run-in-the-Grass, Mutton-Chops, Hedgeheriff, Hayriffe, Barweed, Eriffe, Clite, Clithers, Loveman, Goosebill, Everlasting Friendship, Gooseheriff
USES	Culinary, medicinal, cosmetic
PARTS USED	Whole herb, seeds
HEIGHT	1 to 6 feet, weak, quadrangular or ribbed, prickly, sprawling habit
FLOWER	White, starlike, four petaled, on slender stalks, in clusters of two or three, spring from leaf axils; May to July; tiny, globular, bristly seeds
LEAVES	Lance shaped, narrow, coarse, about ½ inch long, six or eight in whorls along branching stems
ROOT TYPE	Yellowish, thin, branched, fleshy
HABITAT	Rich, moist soil, thickets, hedgerows, edge of woods
PROPAGATION	Seeds
CONTROL	Regular mowing, weeding
CONTRAINDICATIONS	Unsafe for diabetics

leavers is a weed that goes largely unnoticed. Seen in passing, which is how most people see the plant, it looks like one of Nature's scruffier waifs. Too unsteady to stand upright on its own, it forms sprawling tangles of loosely matted greenery on the ground at the base of fences, hedgerows, thickets, or wayside plants. Yet from here, it wastes no time in putting to work the hooked bristles along its stems to hoist itself upward through the vegetation of other plants until it reaches daylight. Once there, it keeps going and frequently manages to blanket its hosts with dense mats of growth. What lends charm to this seeming ingratitude for so much hospitality is that the "blanket" resembles billowy green lace strewn with myriad seed pearls, the plant's tiny white flowers.

This weed always reminds me of a cat I once had who liked to sun herself under the awning of a hedge, her green eyes blinking in lazy recognition, whenever I happened to pass her, and one of her sleek black paws briefly, gently clawing my slacks or shoes. By contrast, cleavers obviously reminded Anglo-Saxons of no such romantic rubbish as cats sunning themselves. In *their* eyes, the weed's waylaying habit more nearly allied it to a *hedge rife*, which in their language meant "a tax collector" or "robber"—a synonymity that seems to have truly withstood the test of time across all cultures and languages.

It goes without saying that, long before either cleavers or the Anglo-Saxons established a foothold in England, the shepherds of ancient Greece were well acquainted with it, as were their flocks. One of the early Greek names for it was *philanthropon*, which may be loosely translated as "loving humankind," and undoubtedly led to the anglicized version, *loveman*, although it seems unfair to the sheep; after all, the weed "loved" *them* even more and in greater numbers.

No wonder Linnaeus eventually gave cleavers the specific name *aparine*, from the Greek *aparo*, "to seize," when he assigned this most common of the bedstraws to the genus *Galium*. Based on the Greek *gala*, for "milk," the name refers to the ability shared by most members of the genus to curdle milk. One member especially, the yellow-flowered *G. verum*, was for centuries used as rennet to curdle milk and as the source of the rich color of England's renowned Cheshire cheese. In fact, although all the *Galiums* are often called "bedstraw," because they were used in former times for stuffing mattresses, the name has special significance in the case of *G. verum*. According to Christian legend, it was one of the "cradle herbs" that made up the hay in the manger at Bethlehem, and so has been called "Lady's bedstraw" ever since.

However, this does not alter the fact that the entire *Galium* genus is relegated among the poor relations of the several plant genera that comprise the vast Rubiaceae family of more than 5,000 species. Star family members are the coffee plant, various medicinal species that furnish quinine and other drugs, and, not least, sweet woodruff, the herb that gives German May wines their perfumed delight.

In spite of its cloying habits—which, to a lesser extent, are shared by the other bedstraws—cleavers can be quite readily brushed or picked off, something that cannot be said for burdock. Also unlike burdock, cleavers is a fairly modest plant, not likely to invade the meticulously tended garden. Of course, neither is it likely to be invited there. On the other hand, in the neglected garden, or in garden beds that are rarely weeded, it "is so troublesome an inhabitant," wrote Culpeper, "that it rampeth upon and is ready to choke whatever grows near it."

On the whole, however, cleavers prefers the sidelines, and is most likely to appear where the blades of a lawn mower cannot or are not intended to reach. Consequently, the weed can often be seen flinging itself wearily over an old tree stump, swooning against a wire fence, entangling some underbrush, or else half-heartedly trying to climb a wooden post, a pole, or the corner of a barn. And it is in such aspects, when its massed flowering peaks in May and June, continuing to bloom in intermittent waves throughout the summer, that cleavers acquires a winsome grace, creating the illusion of feathery clouds skimming the ground.

In these remote points of vantage, its seeds develop inside their bristly, round vessels that are scattered by the wind, or carried off on the clothes of a passing human, on the coat of a dog or cat, or eaten and processed through the digestive systems of grazing animals and delivered by them to new territories. In times past, some farmers used to gather cleavers to give to their poultry; geese have always had a weakness for the weed. Offered half a chance, in fact, geese would probably cause their own extinction, so gluttonously do they devour it whenever the opportunity arises (hence the various "goose" names).

Despite appearances, however, cleavers is anything but frail or useless—or even as lackadaisical as it likes us to believe. On the contrary, although it cannot compete in resourcefulness with such giants of weeddom as the dandelion or stinging nettles, cleavers has a sturdy tradition all its own. It has a respectable reputation to uphold as a delectable food, a superior drink, and a worthwhile herbal medicine.

Its roots produce a red dye, and the herb prepared and drunk as a tea has served as an age-old antiperspirant in Chinese custom. To prepare a deodorant, a strong infusion of cleavers is added to the bath or applied directly to the under-arm. In ancient Greece, according to Dioscorides, shepherds utilized the tangled, rough stems of cleavers as a sieve for milk; some 1,500 years later, Linnaeus was startled to discover exactly the same practice in several remote districts of his native Sweden. And far from least, the seeds of cleavers, when slowly roasted until they are dark brown, and ground, are generally agreed to be a remarkable sub-stitute, a virtual replica, for coffee. And so they *should* be—after all, what's oth-erwise the point in having the right connections?

The whole herb is considered to be rich in vitamin C, and for this reason there is a long tradition in the rural areas of Europe of using the young shoots (for their strong tonic effects) in so-called spring drinks to purify the blood. The expressed juice, taken in twice-daily doses of 3 ounces each, is an old remedy for various skin diseases and eruptions and is said to be beneficial even in cases of psoriasis. For more immediate benefits, many people couple taking the juice with the external application of the crushed plant.

However, because the juice is considered a powerful diuretic, it is not rec-ommended for diabetics. On the other hand, herbalists have found that this diuretic power has proved beneficial in the treatment of edema and of obstructions of the urinary organs particularly, since the plant is said to act as a solvent of stones in the bladder.

According to Culpeper, "The juice dropped into the ears taketh away the pain of them." He also advised that applications of the juice, the powdered, dried herb, or the bruised fresh leaves, would stanch the bleeding of an open wound. In rural sections of France, even now, fresh cleavers is applied as a poultice to varicose ulcers on the legs, as well as to sores and blisters. And Gerard claimed that crushed fresh cleavers is an amazing cure for the bites of snakes and spiders and all sorts of other venomous beasties.

A tea made by pouring 1 pint of boiling water over 1 ounce of the dried herb, infusing this, covered, for 10 to 15 minutes, and drinking it hot or cold, with or without a little honey, is believed to provide a soothing bridge between insomnia and a restful night. Such a tea is also drunk as a mild laxative. And in my grandmother's repertoire of home remedies, it was favored for head colds. And more than once, my grandmother spared me the pain and discomfort of a sunburn by repeatedly splashing a strong cleavers infusion on my roasted skin.

A remedy for scalds and burns already customary in fourteenth-century England was a cleavers ointment that was variously called Heyryt, Cosgres, Clive, and Tongebledes. Undoubtedly, the ointment was prepared in the time-tested country way. This meant that the fresh herb, first "being boiled in hog's grease," to quote Culpeper, would then be strained and stored in a tightly closed jar until needed for colds and swellings. In fact, such an ointment was long ago reputed to have a special curative effect when applied to cancerous growths, especially if the expressed juice of the herb was drunk at the same time.

Although a strong infusion repeatedly splashed on the face is said to remove freckles, cleavers never earned a place on the cosmetics shelf. Its reputation in hygiene, too, seems to be limited to use as a deodorant and antiperspirant. Instead, the weed appears to have concentrated its resources on general health and well-being. Although I myself have not tried this, John Gerard claimed four centuries ago, using Pliny as his source, that "a pottage made of Cleavers, a little mutton and oatmeal is good to cause lankness and keepe from fatnesse." Who knows, it may work.

Modest as are the claims generally made for this weed, even these are amazing, when we consider how weary and listless it is from crawling and climbing and clawing its way through life. It cannot be easy for cleavers to keep blooming and making seeds and making sure that enough people and animals spread them around, while all the time the plant itself is too weak to stand upright.

Yet somehow it overcomes these vicissitudes and finds the strength to be prepared to stimulate and regulate our body's health—if only we'll pay attention. It offers us its very young shoots in spring as either a refreshing, slightly bitter salad, or as a nourishing cooked vegetable. Not only does it taste good but at the same time it is acknowledged to act as a mild diuretic *and* laxative. As if that were not enough, it can also provide us with a terrific cup of after-dinner coffee that's supposed to be better for our health than the stuff we're used to. And when we are ready for bed, doesn't it offer us a special nightcap from the dried leaves that tastes just like black tea but instead helps us get a restful night?

If all this reminds us of somebody, however, there is no need to feel guilty; cleavers know that mum's the word.

Glechoma hederacea

Glechoma hederacea
(GROUND IVY)

VITAL STATISTICS

Perennial

COMMON NAMES	(*Nepeta glechoma, Nepeta hederacea*) Ground Ivy, Alehoof, Field Balm, Earth Ivy, Cat's Foot, Gill-Go-over-the-Ground, Hedgemaids, Gill-Creep-by-Ground, Tunhoof Haymaids, Lizzie-Run-up-the-Hedge, Run-away-Robin, Gill-Go-by-the-Hedge, Robin-Run-in-the-Hedge, Creeping Charlie (or Jenny), Blue Runner, Devil's Candlestick, Hen and Chickens, Rat's Mouth
USES	Culinary, medicinal, cosmetic
PARTS USED	Whole herb
HEIGHT	3 to 10 inches
FLOWER	Purplish blue; sometimes pinkish white; March to July
LEAVES	Scalloped, kidney shaped or roundish, dark green, downy, stalked
ROOT TYPE	Trailing, roots form along square, jointed stalks
HABITAT	Shaded or semishaded areas, roadsides, lawns, woods, under hedges; tolerates sun
PROPAGATION	Root division
CONTROL	Regular cultivation, digging

I have probably known ground ivy all my life, but my first really conscious encounter with it came the day I discovered a silver spoon enmeshed in the weed as I pulled whole clumps of it from an overgrown garden. What a propitious omen, I thought, as I seized the spoon and tossed away the weed. Still, my curiosity was tweaked, not least by the weed's heady balsamic fragrance. Before long, as I began reading about ground ivy, I realized it *had* been a good omen, only I had misread it. Whereas the spoon turned out not to be silver at all, the weed emerged as a veritable vein of gold. Yes, it *is* just a weed—everybody says so—but a weed well worth keeping somewhere in your landscape.

Although both Dioscorides and Galen are credited with having attributed medicinal properties to ground ivy some two millennia ago, we appear to have this information solely on the authority of John Gerard, the sixteenth-century English herbalist and surgeon. At least, his is the first specific mention of ground ivy. Tradition, however, predates this mention by perhaps as much as 1,000 years, to the time of Saxon settlements in Britain. In fact, it may not be too farfetched to suppose that ground ivy crossed the English Channel with the early Saxon invaders. Whether or not they already knew about ground ivy's medicinal properties, it is certain that their use of the plant introduced to England a practice that was to continue until the reign of King Henry VIII. That is, in the days long before hops were introduced, the Saxons clarified their beer by steeping the leaves of ground ivy in the hot brew. This practice simultaneously improved the beer's flavor by imparting to it the characteristic bitterness, and enhanced its keeping qualities by preventing it from turning sour during long sea voyages. After all, the Saxons *were* rovers.

Ground ivy has been and often still is known by numerous common names, most of which evoke delightful images. But what can quickly lead to confusion is the fact that it is among the few plants that may be found listed under two different botanic names. The accepted appellation was catalogued by eighteenth-century Swedish botanist Carolus Linnaeus as *Glechoma*, based on the Greek *glechon*, for "mint" and the Latin *hedera*, for "ivy," which may be loosely translated as "ivylike

mint." But about a century later, along came the English botanist and plant cat-
aloguer George Bentham, who listed the same plant as *Nepeta glechoma*—sometimes
this appears as *Nepeta hederacea*—thereby including ground ivy among the 150
species of the related *Nepeta* mint family. Here, the first of the two names can be
translated as "mint mint" and the second as "ivylike mint." Now, isn't that a
coincidence? By being known as both a *Glechoma* and a *Nepeta*, in a manner of
speaking, it could be said that the ground ivy is its own cousin!

Of the plant's common names, several become self-explanatory with closer
examination; others defy explanation, while the remainder usually refers to one or
another feature of the plant or its growing habit. Only the "gill" names are a little
baffling. *Gill* is said in several sources to derive from the French *guiller*, meaning "to
ferment beer," except nowhere in my French references do I find such a word. The
nearest to it, *gille*, from the Old French and pronounced "jill" in English, used to
refer to "a sort of measure for wine," according to *Webster's*, and has come down to
modern English usage as representing ¼ of a pint. This might be conceivably the
one we need, except that I have seen no indication anywhere that there ever was
such a drink as ground ivy wine. Other gills with a *J* sound used to mean "wench,"
or "lass," or "sweetheart"; in some parts of England it was the dialectal name for
ground ivy—perhaps it still is. Gill-ale also meant ground ivy, or ale infused with
it. Only one of the two gills with a hard *G* makes sense here—the one derived
from the Middle English *gile*, meaning "lip," and this could quite likely refer to the
plant's lipped flowers. In brief, only one thing seems certain: The origin of the
"gill" names is uncertain.

Gerard gives us one of the most succinct, as well as charming, descriptions
of this weed. He writes:

> Ground Ivy is a low or base herbe; it creepeth and spreads upon the ground
> hither and thither all about, with many stalkes of an uncertaine length,
> slender, and like those of the Vine: whereupon grow leaves something
> broad and round; among which come forth the floures gaping like little
> hoods, not unlike to those of Germander, of a purplish blew color: the
> whole plant is of a strong smell and bitter taste. It is found as well in tilled
> as in untilled places, but most commonly in obscure and darke places, upon

banks under hedges, and by the sides of houses. It remaineth greene not only in Summer, but also in Winter at any time of the yeare: it floureth from Aprill till Summer be far spent.

To this, Culpeper's description adds that the plant "shooteth forth roots at the corners of tender jointed stalks," that the leaves are "unevenly dented about the hedges [edges] with round dents," that the "hollow long flowers" are marked "with small white spots upon the lips that hang down," and that the "root is small, with strings."

Although the weed does prefer moist, shaded locations, I have let it roam freely in areas that never receive shade, and still it thrives. On hot, humid days in early summer, when it is walked on, or when the leaves are crushed or picked, ground ivy releases a tart, minty fragrance that is variously described as being reminiscent of citronella, pine woods, or rosemary. The small, translucent flowers attract bees and butterflies and, on occasion, even a hummingbird.

Except in regions where winters are uncommonly harsh, ground ivy keeps its green color throughout the seasons. For this reason alone, it can become a welcome addition to wooded garden areas, in places where the soil is sparse, or as an edging for flower plantings. In this last instance, however, the weed must be kept firmly in check. As a ground cover, it can create almost instant "beds" at the base of trees, or as free-standing "islands," or it can hide an unsightly feature, or be used to prevent soil erosion on an otherwise uninteresting embankment. Few plants can match the rate of growth and density of ground ivy, which is yet another point in its favor. And, of course, all this at no financial cost.

Although some herbalists still prescribe ground ivy for persistent coughs, on the whole, the plant is ignored nowadays. Yet throughout most of its history, its medicinal virtues were extolled for such divers ailments as bronchitis and the "bite of creeping things," for "sore of milt [spleen]," and festering boils. According to Gerard, the Greek physician Dioscorides taught as early as the first century A.D. that "half a dram of the leaves being drunk in foure ounces and a half of faire water for 40 or 50 days together is a remedy against sciatica or ache in the huckle-bone [hipbone]."

In his own time, Gerard writes, "Ground Ivy, Celandine, and Daisies, of each a like quantitie, stamped and strained, and a little sugar and rose water put thereto, and dropped with a feather into the eies, taketh away all manner of inflammation . . . yea although the sight were nigh hand gone." He goes on to say that ground ivy is also "commended against the humming noise and ringing sound of the ears, being put into them, and for them that are hard of hearing."

Culpeper, for his part, declared it to be "a singular herb for all inward wounds, ulcerated lungs, or other parts, either by itself, or boiled with other like herbs; and being drank, in a short time it easeth all griping pains, windy and choleric humours in the stomach, spleen or belly." He also claimed that "if you put to the decoction [of ground ivy] some honey and a little burnt alum, it is excellent good to . . . wash the sores and ulcers in the privy parts of man or woman."

To all of which adds an unnamed old-time writer, clearly a man of few words, "sodden in swine's grease, it is good ointment for all manner aches." This claim is echoed in the traditional belief in ground ivy's cleansing effect on the lungs, kidneys, stomach, and bowels. In fact, according to Dr. Thornton, a nineteenth-century physician, "Ground Ivy was at one time amongst the 'cries' of London, for making a tea to purify the blood." Another nineteenth-century physician flatly hailed the weed as the cure for all lung diseases.

In apparent support of this, modern English herbalist William Smith offers what he describes as "a fine remedy and tonic for the lungs": ½ ounce each ground ivy, comfrey, elecampane, and horehound; ¼ ounce gingerroot, a few grains cayenne (or one or two capsicums), to be simmered in 3 pints of water for 20 minutes, strained into 8 ounces honey or molasses, mixed, stirred, and given as often as necessary in doses of 1 tablespoonful to 1 wineglassful.

An herbal beverage called gill tea was a popular remedy for coughs and consumption among English country people. In fact, one of my sisters and I have discovered that tea made from the fresh or dried leaves is also a pleasant tonic herbal drink—whether or not something ails us. We simply steep 1 teaspoon of leaves in 1 cup boiling water, and sweeten it to taste. In bygone days, such a tea was sweetened with honey, licorice, molasses, or sugar, then cooled and drunk in doses of 1 wineglassful three or four times daily.

Gill tea was also prescribed for whooping cough, bronchial catarrh, and asthma. As an expectorant for bronchitis, a pinch of dried ground ivy infused for

10 minutes in a cupful of boiling milk and strained was taken at bedtime. Because of its reputedly beneficial action on the body's mucous membranes, the tea was also taken for urinary and intestinal complaints. And in America, during the nineteenth century, it was deemed highly effective in the treatment of "painter's colic," or lead poisoning.

As an inhalant, a hot infusion of ground ivy acts as a pleasant relief on head colds and stuffy noses. Snuffing the expressed juice of ground ivy up the nose, for the relief of colds and migraine headaches, is a practice of long tradition, and so is the use of snuff made from the dried, powdered leaves. I can personally vouch that, by merely crushing a handful of the weed and deeply inhaling its balsamic aroma, I have cleared my head many times and deterred headaches, or given myself a quick refresher whenever I have sagged in the midday sun. In fact, the expressed juice of ground ivy was valued already centuries ago for its high vitamin C content in the treatment of scurvy.

An infusion can be used as a lotion, or on compresses, to cleanse sores and ulcers. Combined with the flowers of yarrow and camomile, ground ivy was long considered an excellent poultice for abscesses, boils, and skin tumors. And German herbalists even today treat badly healing wounds with herbal baths of ground ivy.

Although its cosmetic uses are extremely limited, I use a cooled infusion of ground ivy as a most refreshing facial rinse all year round. I apply the warm infusion to my hands when they are chapped. A hot infusion of the plant acts as a stimulating facial steam and can even restore tired feet when they are soaked in it.

However long or short its history may be, however much we may despise it today as just a weed, ground ivy seems always to have found a way to appear unexpectedly, modestly out of the limelight, but ready to serve—just as it appears in our gardens. And so, it is not surprising to find it among the many herbs of Saint John. Known in France as *courroie-de-Saint-Jean*, Saint John's Girdle, ground ivy used to be woven into the crowns worn by dancers around the bonfires celebrating the saint's feast day on June 24.

Matricaria recutita

Matricaria recutita
(CAMOMILE)

VITAL STATISTICS

Annual

COMMON NAMES	*(M. chamomilla)* Camomile, Chamomile, German Camomile, Wild Camomile, Hungarian Camomile, Single Camomile, Mayweed, Scented Mayweed, Pin Heads, Sweet False Camomile
USES	Culinary, medicinal, cosmetic, commercial
PARTS USED	Flower heads
HEIGHT	1 to 3 feet, slender, erect, much-branched, downy stems
FLOWER	Small, daisylike, single flowered, white rays around yellow, hollow disk, fragrant; May to October
LEAVES	Finely divided, feathery, light grayish green, sparse
ROOT TYPE	Thick, fleshy rootstock
HABITAT	Roadsides, waste places, fallow pastures, grain fields
PROPAGATION	Seeds
CONTROL	Cultivation, weeding

Camomile is a cheery plant. Its sole aim in life is to heal, charm, soothe, and refresh. It is an herb that has been known for thousands of years and prescribed by physicians since medicine was first developed. It is native to Europe, North Africa, and the temperate regions of Asia. Yet for centuries, perhaps always, this modest plant has had a rival—another camomile. Which of them is the *true* camomile? Which of them will win the Princess and the Pea award? Debate and confusion will likely continue into the next millennium.

Linnaeus named it *Matricaria chamomilla*. Because it has always been particularly valued by Germanic peoples, it came to be known as German camomile. Essentially, it has remained a wild plant, an annual, although it can be cultivated and often is. Recently, botanists renamed it *Matricaria recutita*.

Linnaeus named it *Anthemis nobilis*. It is perennial, prefers moist, temperate climates, and is usually cultivated. Known as English or Roman camomile, it is mostly associated with England, where it is grown into fragrant lawns. Recently, botanists renamed it *Chamaemelum nobile*.

The supporters of both plants staunchly and adamantly refer to their respective candidate as simply "camomile"—which is often spelled "chamomile"—or as *true* camomile. Sometimes, as is not uncommon in rivalries, they even feel driven to denigrate "the other." A few sources skirt the entire issue by simply ignoring both plants; some make the effort to treat each one fairly, squarely, and separately; others point out the plants' few differences, then proceed to treat them as one. And a few discuss their candidate so exclusively that the other camomile might as well not even exist.

Luckily, both plants are considered equally benevolent and equally prepared to serve us. They both owe the name camomile to the lively applelike fragrance of their similar, daisylike flowers. Loosely translated, the name means "ground apple," derived from the Greek *chamai*, for "ground, soil, or earth," and *melon*, for "apple." Similar, too, is their rather feathery foliage. The strong-scented flowers and foliage of both plants contain a bitter medicinal principle, which is used in aromatic bitters. The chemical properties of both camomiles are sufficiently alike to make the "blindfold test" pointless.

The greatest differences between the plants are physical and visible. German camomile, an annual, may reach a height of 2 to 3 feet. Upright and much branched, it bears its terminal flowers on numerous stalks. The flowers are single, composed of white rays surrounding the central yellow, *hollow* disk. Roman camomile, a perennial, is a creeping, trailing plant whose wispy, feathery leaves and stalked flowers rarely grow above 10 inches. The flowers are somewhat larger than those of German camomile, and their central yellow disk is broader, as well as *solid*.

Although the earliest recorded use of camomile, so far, goes back some 5,000 years, it is impossible to say with certainty which of the camomiles it was. Perhaps it was both. There is indication that *M. recutita* is indigenous to Europe and western Asia, whereas *C. nobile* is indigenous to southern and western Europe. My decision to include *M. recutita*, instead of "the other," in these pages is based on the fact that it is more prevalent as a wild plant and is generally considered a weed. Herbalists on both sides of the Atlantic say that the two plants can be used interchangeably. For all these reasons, I do not hesitate to claim for German camomile any historic or folkloric reference that may appear elsewhere as pertaining only to English or Roman camomile.

As is true in modern medical practice, ancient Egyptians placed great emphasis on hygiene and diet. Oil of camomile was one of the numerous drugs known to them, which they prepared as ointments, potions, and poultices. By 3,000 B.C., physicians were regarded with so much awe that they were deified. One such "god" was Imhotep, the very real and earthly physician to King Zoser.

Camomile was regarded as sacred, too. "Thys herbe was consecrated by the wyse men of Egypt unto the Sonne and was rekened to be the only remedy of all agues," wrote William Turner, an English botanist, in the sixteenth century. A century later, Culpeper was clearly unconvinced. The Egyptians "were like enough to do it," he observed caustically, "for they were the arrantest apes in their religion I ever read of."

Early Germanic tribes dedicated the plant to Balder, their god of light and peace. The Greeks and Romans instead viewed camomile with a more practical eye. Both Dioscorides and Pliny prescribed its use in bath or poultice for ailments of the liver, kidneys, and bladder, as well as for headaches. They also treated fevers and female disorders with it.

Describing its appearance as well as its power, Turner noted that camomile "hath floures wonderfully shynynge yellow and resemblynge the appell of an eye. . . . It will restore a man to hys color shortly yf a man after the longe use of the bathe drynke of it after he is come forthe oute of the bathe." Turner's spelling may lack the modern penchant for brevity, but he makes his point. Even Culpeper, having first denigrated the Egyptians' belief that camomile could cure all agues, proceeded to offer an impressive list of medicinal successes for the plant.

On the other side of the English channel, the well-known sixteenth-century German herbalist Hieronymus Bock claimed that no more useful medicinal herb exists than camomile. Several of his peers agreed, adding that the herb relaxes tense muscles and soothes swellings. Much later, Thomas Panckow, physician to the Elector of Brandenburg, recommended a vigorous back rub with camomile oil to alleviate fatigue. He also observed that a lizard that has been wounded by a snake heals and recovers by rolling in camomile plants.

Earlier this century, Johann Künzle, known as the Swiss *Kräuterpfarrer* (herb priest), recorded some remarkable cures effected by Beth, the "camomile witch." With camomile massages, it seems, she restored the strength in lame or weakened limbs; she healed inflammations of the eyes with warm compresses of camomile simmered in milk; she cured urine retention with small cupful doses of camomile decocted in wine; she effected a complete recovery from rheumatism with the twofold treatment of massaging the affected limbs and having the patient drink a small glassful of wine in which camomile was steeped overnight; to rid yet another patient of a severe head cold, Beth ordered her patient to wash his hair with warm camomile tea three to five times daily.

Camomile has always been considered an excellent remedy for children's ailments. Russian infants suffering from constipation used to be given 1 to 2 teaspoons of camomile tea; the same dose was used in many countries to relieve babies suffering from flatulent colic and the pains of teething. The prescription I like best comes in what sounds like the unmistakably world-weary voice of an old-time country doctor who has seen it all, who is shocked by nothing, and who has infinite faith in the force of life. He suggests an infusion of 6 to 8 camomile flowers in a teacupful of boiling water. Although he fails to mention it—old-time doctors took common sense for granted—the tea should cool before being administered. "Two teaspoonfulls of such an infusion will quickly quiet the howling child with gas pains," he advises, adding that "simple gas disappears from little ones as if by magic."

Having been subject to bladder infections as a child, I know of camomile's healing powers from personal experience. Because we lived for a time in the part of Germany that had spawned Father Sebastian Kneipp, Mother routinely treated all our childhood ailments according to his renowned water cures. At the first sign of trouble, she ordered me to bed. Across my lower abdomen she placed a small towel that she had dipped in a strong, warm camomile infusion and wrung out. Over the wet towel, she laid a large dry one, then tucked me under plenty of warm blankets, and left me literally to sweat out the infection. After an hour or so, she completed the treatment with a cool wash, followed by a vigorous rubdown with a dry terry towel. Then she had me drink a cup of warm, honey-sweetened camomile tea and left me to rest or sleep another hour. Only rarely was it necessary to repeat the treatment.

Modern herbalists regard camomile as antiinflammatory, antispasmodic, digestive, nervine, sedative, diaphoretic, as well as a tonic and a powerful antiseptic. The plant is rich in potassium and calcium. It is considered beneficial not only to man but to beast as well, internally and externally. Camomile even influences the well-being of its own kind, other plants. Dispersed throughout a garden, this "plant's physician" is said to be capable of restoring the health of any sickly plants nearby. Camomile tea sprayed on seedlings prevents damping off or mildew; camomile plants added to the compost heap aid in the breakdown of vegetable matter. Small numbers of the plant in a wheat field encourage good growth of the grain; in large numbers, camomile impedes the grain's development.

Midday is the ideal time to gather camomile blossoms, when the power of their essential oil reaches its peak. (The same is generally true of all herbs.) Only the freshest camomile blooms should be selected for use, just before the white rays of these have begun to recurve (turn down), at which point the flowers resemble miniature shuttlecocks. When I pinch or cut off the flowers, I try to avoid including parts of the stems, as these add a bitter flavor. Fresh, they are best used immediately, before they wilt; otherwise, I loosely spread them to dry on brown paper in a dry, airy attic or room, or in an oven on low heat.

Of course, you may think all that bother seems unnecessary in this instance. After all, so well known, so popular is camomile tea, that no self-respecting supermarket or corner grocery store would dream of not stocking it.

In much of the world, camomile tea is considered an excellent remedy for nervous disorders, diarrhea, and flatulence. Most often, people drink a cupful of camomile tea at bedtime to assure themselves of a restful night's sleep. By some it is believed both to prevent and to cure nightmares. Herbalists recommend it as a diuretic, particularly in instances of edematous swellings. It also has a long tradition of stimulating the appetite, of alleviating indigestion and heartburn, and of being helpful for gout. It has even been credited with arresting an attack of delirium tremens in the early stage.

A cup of camomile tea is recommended as an appetizing tonic, especially for the aged, if it is drunk about an hour before a meal. The tea is also prescribed for diarrhea, and camomile is taken to ease the pains of arthritis and to soothe conditions that are characterized by heat flare-up, redness, and swelling. Through the centuries, a strong camomile tea, with or without a little fresh lemon juice, has been prescribed for the quick relief of migraines and fevers, menstrual cramps, the aches of flu, and neuralgia.

German herbalists recommend one wineglassful of "camomile wine," to be taken mornings and evenings after meals to relieve urine retention. To prepare such a wine, according to Treben, 4 heaping teaspoons of fresh or dried camomile blossoms are placed in 1 liter white wine, which is heated to near boiling, covered, and infused for 10 minutes, strained, and bottled.

Externally, a poultice of crushed fresh camomile flowers alone, or in equal proportions with crushed poppy heads, can be applied to heal swellings, inflammations, abscesses, and sores. As a gentle wash, a strong, lukewam infusion is used to cleanse wounds and sores, and to soften the skin around them. A compress dipped in a warm infusion and applied externally is a long-used remedy for a toothache, inflamed eyelids, and the pains of neuralgia. Another well-known

German method for allaying neuralgia is to hold a warmed herbal cushion to the affected part. This is also considered an excellent treatment for leg cramps. A simple herbal cushion can be made by sewing dried camomile flowers between two small handkerchiefs, for example.

Used as a steam inhalant, a camomile infusion is recommended for the relief of congestion resulting from colds and catarrhs, of inflamed mucous membranes, and of infection of the respiratory tract. Based on German herbalist tradition, for such an inhalant, I pour 1 quart boiled water over 2 ounces fresh or dried flowers in a large bowl. With a towel thrown over my head and the bowl, I deeply inhale the steam for 15 to 20 minutes, then gently dry my face, and rest 10 to 15 minutes. If the skin around my nose and mouth is reddened and tender, I next apply a little camomile (or other herbal) oil.

A strong decoction added to bathwater is said to ease rheumatic pains and is considered a soothing treatment for hemorrhoids. A camomile bath is also wonderfully relaxing at the end of a long day, even if nothing ails you. Place 3 ounces flowers in 1 quart cold water, cover, and bring to *near* boil. (Never boil camomile, as this destroys the valuable properties contained in its oils.) Immediately remove from heat, and, still covered, allow to infuse for 20 to 30 minutes before adding to bathwater. If you plan to be outdoors later on, save some of the decoction and apply it to all exposed skin to prevent insect bites.

In many European households, probably no other camomile preparation is so much relied on for its healing effects as camomile oil. There it is massaged warm into painful joints and limbs to alleviate muscular cramps and sprains and to soften weather-roughened skin. Or, applied warm to a compress around the throat, it is used to restore loss of voice. Not least, skin infections, scabbing, and eczema are treated with applications of the oil.

In one of the commonest methods of preparing the oil, 1 pint olive oil is poured over 3 ounces flowers in a 1-quart jar. Covered, the jar is set in the sun for 3 days and lightly shaken once daily. The contents are next emptied into the top of an enameled, glass, or stainless steel double boiler, covered, and allowed to macerate for 2 hours over low heat, occasionally stirred, and finally strained and pressed through muslin. The remaining flowers can be used as a poultice for muscular cramps, rheumatic pains, or swellings, and even for the relief of piles.

In cosmetic usage, camomile is capable of providing a complete beauty treatment. I make my own skin-cleansing oil as above, but use sunflower or almond instead of olive oil (safflower oil or wheat germ oil can be substituted). To remove blackheads, an old English herbal method is to soak 2 tablespoons flowers in 1 cup cold milk overnight, then warm it slightly before bathing the face with it. This preparation is also considered an excellent lotion for roughened hands. In my own opinion, nothing quite equals the gently soothing effects of a camomile facial steam. For a favorite facial mask, I particularly like to use 2 teaspoons camomile infusion mixed with ½ teaspoon honey and 1 teaspoon yogurt. Afterward, I rinse my face with chilled camomile infusion, which acts as a mild astringent. For my eyes I use compresses wetted in a chilled camomile infusion (this also helps refresh tired or puffy eyes). I can even use camomile tea for a refreshing mouthwash.

A camomile infusion used to shampoo and rinse hair not only brings out golden highlights in blond, red, or light brown hair but also softens *all* hair. One of my sisters swears by it. After a bath, she also splashes herself generously with camomile water. She makes this by pouring 1 pint boiled water in a 1-pint jar filled with fresh or dried flowers, letting them cool 1 hour, then adding 1 teaspoon rubbing alcohol. She strains the herbal water, then bottles and refrigerates it, ready for use.

With the bulk of its existence devoted to healing and beautifying the human race, camomile has found little time to come near the kitchen or dining room, except in a remedial capacity, as a tea, or, formerly, in herb beers. Nevertheless, it is very likely that many an epicurean palate has been tickled by a glass of manzanilla, the light Spanish sherry named for and flavored with the "little apple," which means camomile. And in Paris in times past, camomile flowers boiled with orange peel used to make a refreshing "water for washing the hands at table," according to *The Goodman of Paris*, the late-medieval book of household hints. Downstairs, in the meantime, who knows how many cooks across Europe were soaking large quantities of meat in camomile tea to mask or eliminate the odor of rot. Before refrigeration, it seems, this was common practice.

Melilotus officinalis

Melilotus officinalis
(MELILOT)

Annual

COMMON NAMES	Melilot, Sweet Clover, Yellow Sweet Clover, King's Clover, Moonseed, Sweet Lucerne, Corn Melilot, Plaster Clover, Wild Laburnum, Hart's Tree, Hart's Clover
USES	Culinary, medicinal, cosmetic, commercial, household
PARTS USED	Whole herb
HEIGHT	2 to 5 feet; much-branched, smooth, erect stems
FLOWER	Yellow, fragrant; in slender, tapering, terminal spikes; May to October; black, wrinkled, one-seeded pods
LEAVES	Elongated, trifoliate, smooth, slightly toothed, fragrant
ROOT TYPE	Large, spreading, woody, white root
HABITAT	Fields, waste places, roadsides, hedgerows
PROPAGATION	Seeds
CONTROL	Regular cultivation

There was a time when the melilots, a group of weedy plants also known as sweet clover, were assigned to the genus *Trifolium*, together with the red and white clovers. Eventually, however, they came to be classified separately, as the genus *Melilotus*. The name stems from the Latin *mel*, for "honey," and *lotus*, probably for a legendary "fruit said to induce a dreamy indolence and forgetfulness." At least, I like to think so. Certainly, bees favor these weeds; in fact, after observing them swarming and swirling around melilots in bloom, I would go so far as to say that bees become positively intoxicated by the sweet fragrance of the plants.

Of the roughly twenty species in the genus *Melilotus*, only three are generally known in the United States. They are the yellow-flowered annual, *M. officinalis*, the white-flowered biennial, *M. alba*, and the perennial, *M. indica*, also yellow flowered. They also include a relatively "new" species, the *annual* variation of *M. alba*, which was discovered some decades ago, and is known as Hubam clover, for Professor Hughes, its discoverer, and for Alabama, its native state. Following the tests to which it was subjected, this foundling quickly won wide recognition, both for its general adaptability to varying climatic and soil conditions and for its abundant production of nectar and seed. Indeed, some experts believe there is every reason to think that this modest weed will, in time, emerge as the world's chief honey-producing plant. (The blue melilot, used in the manufacture of *Schabzieger*, a particular kind of Swiss cheese, belongs to the related pea-family genus *Trigonella*.)

In appearance, especially at a quick glance, the melilots are very similar to alfalfa, but can be just as quickly distinguished by their color. Whereas alfalfa is of a blue-green hue, the melilots are a yellowish green. All melilots are considered weeds and exude a vanillalike fragrance that intensifies when the plant is dried. And all the melilots are of southeastern European origins, from around the Mediterranean Basin and Asia Minor. Very likely they introduced themselves into the United States by means of their seeds clinging to the backs of sheep brought here by early immigrants. Today, the weed is a familiar sight throughout much of this country.

There is no denying a close relationship between the melilots and the clovers. Together with such varied plants as alfalfa, broom, and some locust trees, as well as the peas and green beans we eat, they are members of the large legume or pea family. This remarkable clan is considered the most useful in nature, because, rather than taking nitrogen from the soil like other plants, the legumes give nitrogen *to* the soil. The melilot, like all the legumes, does this by obtaining its nitrogen supply from the air and feeding it to the millions of special bacteria that live in nodules on the plant's roots. These bacteria have only one purpose in life—and a short life it is—which is to fix the nitrogen and give it to the plant. Through this ingenious device, melilots are able to ensure their own growth while leaving untouched the nitrogen in the soil. Even when they die and are turned under, the plants bequeath to the earth both their nitrogen supply and the excellent humus derived from their leaf decay.

Not surprisingly, their nitrogen-fixing capability has placed the melilots among the most valued soil-building weeds. Under the name sweet clover, their seeds can be purchased by the pound from agricultural suppliers and sown as field-cover crops, to smother undesirable weeds, to protect land from erosion, to improve soil—or to produce green manure in the private garden. Able to live even in unfavorable conditions, the white-flowered biennial melilot, which is sometimes known as "Bokhara" clover, is often used for reclaiming poor or eroded land, and for protecting the soil on steep, denuded slopes.

In former times, melilot was extensively cultivated as a forage crop, especially in England, where landholders viewed it as excellent pasture and hay. Today, the practice is less widespread, and it is mostly the Hubam clover that is grown for the purpose, because it matures in a single season. The opinion of cattle and horses through the centuries, however, has tended to be a little ambivalent. To them, fresh melilot is an acquired taste; however, they do enjoy a good munch of melilot hay. Deer, on the other hand, have always browsed contentedly on the fresh weed, hence the common names hart's tree or hart's clover. But then, show me a deer that does *not* browse contentedly on almost anything growing in its path!

Melilot hay became the unlikely source of an important scientific discovery in the 1920s, when American farmers began to store bales of sweet clover for fodder. To their utter consternation, the cattle feeding on this hay began to hemorrhage internally. The immediate cause was found almost at once—the hay, having been stored incompletely dried, had fermented. Yet this fact did not fully account for such dramatic consequences, because it is not uncommon for hay to be stored before it is entirely dry, nor for it to enter a fermentation process as a result. Further investigation soon revealed a Jekyll-and-Hyde quality in the melilot's chemical composition. It seems that coumarin, the property that gives the weed its delicious vanilla fragrance and flavor, turns into the anticoagulant dicumarol when the plant is fermented. Yet unrotted melilot is completely safe.

Except for the color of their flowers, yellow or white, there is little difference between the various species of melilot, nor in the uses to which they can be put. Amazingly enough, although even the kindest souls do not shrink from calling them weeds, the melilots have a long and respected tradition of service not only out in the fields but domestically as well, as medicine, food, and even perfume.

For medicinal purposes, the entire herb, including the flowering spikes, has always been gathered in early spring to be used fresh or dried. Dried, the herb's fragrance becomes greatly intensified, whereas its flavor remains slightly bitter. In addition to its main constituent, coumarin, melilot is also a source of protein. The plant is credited with having mildly astringent properties, for which it has been traditionally considered of value in poultices to treat inflammations and wounds of the eyes and other tender parts of the body. Taken as a tea, it has a long tradition as a digestive and mild diuretic. Applied externally, it has been respected throughout history as an emollient and a treatment for relieving arthritic pains.

The early Egyptians treated earache and intestinal worms with a tea made from melilot, and Anglo-Saxons believed in the plant's ability to preserve eyesight. The great physician Galen, of ancient Athens, even then applied melilot plasters to the swollen joints and inflamed tumors of his imperial and aristocratic patients. In Greece, so Nicholas Culpeper tells us, plasters often consisted of herbs heated and stirred together with powdered minerals and fatty substances into a stiff consistency. "Then they made it into rolls, which when they needed for use, they could melt it by the fire again."

In the sixteenth century, however, according to Gerard's writings, the preparation of a plaster seems to have been rather cumbersome. "Melilote boiled in sweet wine untile it be soft," Gerard directs, "if you adde thereto the yolke of a rosted egge, the meale of Linseed, the roots of Marsh Mallowes and hogs greeace stamped together, and used as a pultis or cataplasma, plaisterwise, doth asswge and soften all manner of swellings."

Today, such a plaster might be prepared by mixing the dried, crushed herb (possibly a mixture of herbs) and a little water, spreading the paste on cotton cloth, and placing this on another cotton cloth laid over the affected area. Today, too, a not dissimilar remedy is prepared in some parts of Europe, as it was in Galen's day, using melilot, olive oil, resin, and wax. Indeed, just as melilot ointment continues to be applied to wounds and sores, so melilot plasters are still available from some English pharmacists.

Still recommending melilot, Gerard also maintained that, mixed with wine, "it mitigateth the paine of the eares and taketh away the paine of the head." For the same purpose, instead of using wine, Culpeper suggested steeping melilot in vinegar or rose water. He also preferred poultices made of the fresh plant, to treat not only hard swellings and inflammatory tumors but all inflammations. He further credited melilot with easing stomach pains when applied either fresh or boiled with "the yolk of a roasted egg, or fine flour, or poppy seed, or endive," whereas the fresh juice dropped into the eyes "is a singular good medicine to take away the film that dims the sight." For people who are subject to swooning, he recommended frequently washing the hair with the distilled water of the herb and flowers. This, he wrote, also strengthens the memory and comforts the brain. If such a claim seems a little farfetched, a seventeenth-century recipe even recommends a "bath for melancholy" in which melilot is one of several ingredients.

Idly stripping ripe melilot seeds from their spikes in late summer is hardly apt to spark thoughts of cookies and puddings or soups and salads during the average person's stroll down a country road. Yet these are some of the edibles in which sweet clover has made a gustatory difference over the centuries. Of course, by summer one will have missed the weed's most versatile contributions to the palate, until next spring. Even so, the seeds, which look like small peas, can add a delightfully different flavor to stews and soups, especially lentil, bean, and pea soups. The seeds are used whole or ground into powder, and can be easily dried and stored. Sprinkling them sparingly as a powdered seasoning, I enjoy the understated combination of pungency and vanilla fragrance they add to familiar preparations of meats, potatoes, and pasta.

Among nature lovers and foragers, the fresh young leaves in spring, before the flowers begin to blossom, are particularly pleasant additions to salads. The leaves can also be prepared as a vegetable, steamed or boiled for 3 to 5 minutes and served simply with butter and lemon juice to accompany fish or chicken. But it is the crushed, dried young leaves that positively beg to be added to cookies, cakes, pastries, pies, fillings, fruits, and all the other goodies for which recipes prescribe vanilla. And an infusion made from the leaves alone, or mixed with flowers, can not only be taken for the relief of flatulence but for the simple pleasure of a refreshing drink.

Although the distilled water made from the fresh flowers has been for flavoring foods, more often, the flowers have found their way into the preparation of a simple, pleasing toilet water. Such waters are easy to make with a wide selection of fragrant herbs, singly or blended, and provide a deliciously cooling splash on a steamy, hot summer day. With melilot, my own method is to soak ½ ounce dried melilot leaves in 4 ounces vodka for one week. Over 2 tablespoons each of crushed fresh lemon verbena and orange mint, and 1 teaspoon fresh rosemary I pour 6 ounces boiling water and steep, covered, 20 minutes. Then I strain both liquids through double muslin, combining them in a tightly closed bottle.

For a change from more familiar fragrances, or to avoid the overpowering smell of camphor, I also place sachets of dried melilot among my clothes and linens. Not only does the dried herb add its pleasant scent to these but, more important, it helps protect my clothes and linens against moths. Or, depending on the available space, you might prefer to place dried melilot sprigs on shelves, in drawers, or on the floor, or hang them on closet walls. For one of my own favorite moth repellents, I blend dried melilot and peppermint. I fill numerous sachets with the mixture and place them in coat pockets, between sweaters, two or three in each clothes drawer, and on clothes hangers. By simply crushing the sachets occasionally, their fragrance and effectiveness are renewed indefinitely.

All in all, melilot is a plant with much to recommend it. It is a threefold marketable commodity—for honey, for fodder, and for soil improvement; it is a food and a medicine, nature's own soil protector, minor household help, money-saving flavoring agent, pleasant perfume, even fragrant snuff. So why is this weed virtually unknown in general usage? Why does the name melilot mean nothing at all to most people, and the name sweet clover bring to mind only images of cousin red clover? Most of all, why is it considered a weed at all? Surely, in the course of time, it has proved beyond doubt that it has never qualified for weedhood, that it long ago earned its place among the many herbs and wildflowers we admire.

Nasturtium officinale

Nasturtium officinale
(W A T E R C R E S S)

VITAL STATISTICS

Perennial, Aquatic

COMMON NAME	Watercress
USES	Culinary, medicinal, cosmetic, commercial
PARTS USED	Leaves, above-water stems, flowers
HEIGHT	To 12 inches; fleshy, hollow, branching, floating stems
FLOWER	Tiny, white, four petaled, in elongated clusters; April to October
LEAVES	Divided into three to nine dark green, oval, heart-shaped leaflets; terminal leaflets larger than others
ROOT TYPE	Fibrous, along creeping, floating stems
HABITAT	Springs, brooks, running water, freshwater ponds
PROPAGATION	Seeds, cuttings
CONTROL	Weeding, eating

Watercress. Not for a moment does this simple name even hint at the botanical convolutions that enmesh it. Like the cresses as a whole—most of them regarded as weeds—watercress is a member of the enormous Cruciferae, the Mustard family. Within this clan, it is the only species belonging to the genus *Nasturtium*, a name derived from the Latin *nasus*, for "nose," and *tortum*, for "twist," in allusion to the plant's pungent, nose-twisting flavor. The specific name, *officinale*, indicates a medicinal value.

Not altogether surprisingly, the plant's botanic name suggests a kinship with a popular, usually trailing, garden plant. However, there is absolutely no truth in the rumor that *N. officinale* is in any way related to the garden nasturtium, whose botanic name is actually *Tropaeolum majus*. The fact that *T. majus* is also popularly known as Indian cress, although it is not a cress at all, is strictly a coincidence. And the fact that it happens to be equally pungent and wholesome as watercress merely proves that a plant does not have to be a cress to taste like one.

Just when the average individual has digested all this genealogy, there comes more. In some older references, the botanic name for watercress may appear as *Roripa nasturtium-aquaticum*, or as *Radicula nasturtium-aquaticum*. Not to worry; Gerard referred to it as *Nasturtium aquaticum minus*; in Latin, he wrote, it was called *Flos cuculi*, "in English, Cuckow-Floures," because it blooms "in Aprill and May, when the Cuckow begins to sing her pleasant notes without stammering." Culpeper called it *Sisymbrium nasturtium aquatica*. At least, everybody seems to have agreed on the nose-twisting aspect of the plant.

By now, you could be forgiven for muttering that all this information is "not worth a curse," an expression that predates Rhett Butler's parting shot to Scarlett. Certainly, you will be forgiven for not connecting *curse* with *kers*, which is an old English word for cress, based on *cerse*, the name the Anglo-Saxons gave to the plant they used to eat both raw and boiled.

As is true of so many plants, watercress immigrated from the Old World, together with early settlers. Although it was cultivated at first, being a restless plant it soon escaped to the wild. Rapidly, it claimed a wide range of territories of its own, always seeking out cool, clear waters. Today, it is a naturalized wildling over most of the temperate and frigid regions of North America, washed in the shallows of brooks and springs, stretched out floating in the fresh streams that flow through fields and marshes. Yet, in spite of its independence, watercress rarely strays too far from human habitation, and is, in fact, cultivated for sale in some areas—for good reason.

Watercress is said to have an exceptionally high content of iodine and iron, as well as vitamins A, B$_2$, C, D, and E, and the minerals calcium, copper, potassium, and magnesium. In short, watercress is a storehouse of valuable component elements, and these are at their best when the plant is in bloom. It is prescribed to treat diabetes and anemia, skin disorders like eczema, poor night vision, and to strengthen children's teeth and bones. It used to be considered a specific for tuberculosis and scurvy. In modern times, its freshly expressed juice in spring forms a popular tonic drink in Europe for cleansing the blood and stimulating the appetite. Herbalists prescribe it as an antiscorbutic and a febrifuge, and regard it as useful for treating rheumatic pains, nervousness, and bronchial catarrh. In different lands, it has been variously relied on as a laxative and in the treatment of asthma—even as a contraceptive.

Writing about 400 B.C., the Greek historian Xenophon noted in *Anabasis*, his famous record of a military expedition into Persia, that he had encouraged Persians to feed watercress to their children to help them grow strong. Ancient Greeks and Romans believed in the power of watercress to sharpen the mind. By the first century A.D., such noted Roman poets as Ovid, Martial, and Columella were testifying to its erotic power. They called the plant *impudica*, or shameless. In fairness, it is uncertain whether they meant watercress or one of the field cresses. What *is* certain is the long tradition in the East, even then, of cultivating cress as an aphrodisiac. Apicius, a Roman contemporary of the poets and himself the author of a cookbook, recommended cooking onions in water together with pine seeds or cress juice and pepper. This, he claimed, would lead without fail to the consummation of desire. The cure for impotence, according to the Roman physician Marcellus Empiricus, could be trusted to his prescription of cress, red onion, pine seed, and Indian nard (spikenard), to be taken in small, equal proportions.

Viewing watercress through more prosaic eyes, Europeans during the Middle

Ages considered it a valuable ingredient in ointments for healing wounds inflicted by swords and lances. In France, it is said, Napoleon I became so fond of watercress that he had it planted around the frost-free springs at Nîmes, so that he would not have to deny himself the pleasure of the plant even in winter. Who knows, of course, whether he might not also have heard of its reputed aphrodisiac propensities. In the New World, once watercress had settled down along numerous stream beds, American Indians put it to work as a treatment for kidney and liver ailments. And, like people everywhere, they also adopted it as a food.

Wherever it grows along flowing water, watercress is as likely as not to disappear one day, without warning, in much the same way it may have arrived. High winds may rip it loose and set it adrift, or floodwaters in spring may sweep it away. If you follow the stream, you may be able to find the cress a mile or more downstream, nestling in a sunny curve, or clinging to the low-hanging branch of a shrub along the bank of the stream.

If you have some running fresh water of your own, pull up a handful of watercress (be sure it comes from water that is free of contaminants). Set it into the silt of your own stream or spring-fed pool, away from the mainstream. Before you know it, you will have an ever-expanding patch of your own watercress. On the other hand, if you prefer to start from the beginning, the seeds are easy enough to obtain through seed catalogues. Soak them overnight, then drop them in the silt at the water's edge. Nature will take care of the rest.

Watercress is delicious however it is served, especially fresh, and especially in May and June, when its active principles are at their best. The leaves and stems are most often used, although the flowers are also edible. The flavor of watercress is peppery, with a hint of mustard, tangy, but not too sharp. However, most herbalists caution that *excessive* consumption of watercress may lead to bladder troubles for some individuals.

The most essential aspect of gathering watercress is to do so from a stream that is clean. It should not be gathered from streams that might be contaminated by the droppings of sheep and other livestock, or that might contain harmful chemical wastes. The stems should always be pinched or cut at or above water level (below the surface, the stems are tough).

Once the leaf stalks are trimmed from the main stems, the latter can be set aside for later inclusion in watercress soup. Rinsed and patted dry, the leaves can be added raw, whole or chopped, to sandwiches and to almost any kind of salad, be this hot, cold, or jellied. Cress is used as garnish on cheeses, meats, eggs, and vegetables. Or, quickly tossed with sliced mushrooms in sesame oil over a high flame, enough to wilt slightly, watercress can be served with grilled steak or braised chicken breast. I like to prepare a piquant herb butter for grilled meats or seafood by finely chopping the leaves and blending them with onion and a little grated lemon peel into lightly salted butter. Or, by mixing them with finely chopped onion and pimiento, blended into cream cheese, I prepare a zesty spread for hors d'oeuvres. I also use the spread as a filling for omelets. Finely chopped watercress leaves and tops can also be added to hollandaise sauce, to be served over freshwater fish.

By far my own favorite use of the plant, apart from salads, is in watercress soup. How often Mother made this for us when we were children, especially on Sundays when we returned from our favorite walk! This always led us to the end of a narrow, meandering valley, to a long-disused mill-turned-inn, where we could get a glass of lemonade. Darting from one side of the path to the other, sometimes disappearing for a while in one of the fields, was a brook of sparkling clear water. Drifting and rolling on that water, amid the grasses and flowers that overhung the edge, were dense cushions of watercress. On our way home, we often picked enough for at least two helpings each.

Probably everybody who has ever encountered watercress has either developed or discovered his or her own "most delicious" recipe for it. Inevitably, I have mine. It makes six servings. Slice 4 to 5 scallions and 1 medium onion and sauté these in 3 tablespoons butter until wilted. Add 4 medium-sized peeled and sliced potatoes and 1 quart chicken stock. Bring to a boil, then immediately reduce heat and simmer, covered, 30 minutes. Separately, drop a generous handful of watercress into a little boiling water, cover, boil 1 minute, and strain. Purée the soup stock in a blender, adding watercress. Return soup to pan, add 1 cup milk, heat, and season to taste with pepper and salt. Stir in 1 cup light cream, or yogurt, just before serving hot or chilled. Garnish with watercress leaves.

Mother's recipe was simpler; it was wartime. In a little margarine (say, 3 tablespoons), she sautéed the rinsed and chopped watercress. When this was wilted, she stirred in 3 tablespoons flour, cooking it over low heat to a light, nutty color and aroma. Then, all at once, she added 1 quart hot stock (or water), stirring the soup vigorously until it was smooth, and seasoned it with pepper and salt. Sometimes, she added half stock, half milk, yet it was every bit as satisfying as my own much later version.

Apart from its tangy flavor, what makes eating watercress all the more satisfying in my view is the knowledge that every mouthful also helps to keep me healthy. Eaten raw, it not only prevents inflamed or bleeding gums but is considered one of the best natural depuratives. Whether it is eaten regularly in salads, as garnish or soup, or whether it is drunk in an infusion, or as freshly expressed juice mixed with an equal amount of milk or cold chicken broth, watercress is recommended for its remineralizing action and its value in treating nervous debility, urinary disorders, biliary complaints, bronchitis, and edema.

Watercress juice or the crushed leaves dabbed on the skin every day is said to remove facial blemishes; applied under the arms, they are known to be of service as a deodorant, particularly in the case of individuals who eat lots of fresh vegetables and fruits.

With so much to recommend watercress, no wonder Culpeper's attitude toward nonbelievers was rather peremptory when he wrote, "those that would live in health may use it if they please, if they will not, I cannot help it."

P S : Having introduced the garden nasturtium—you remember, *T. majus*—I feel somehow honor-bound to mention that every part of that plant—leaves, stems, and flowers—can be put to all the same uses as watercress, plus one more: The "capers" made from the seedpods are even better than capers!

Phytolacca americana

Phytolacca americana
(P O K E W E E D)

VITAL STATISTICS

Perennial

COMMON NAMES	(*Phytolacca decandra*) Pokeweed, Poke, Pokeberry, Inkberry, Virginia Poke, Scoke, Red-Ink Plant, Garget, Pocan Bush, Pigeonberry, American Cancer, American Nightshade, Chongras, Cancer Jalap, Coakum, Bear's Grape, American Spinach, Cancer Root, *Raisin d'Amérique*, Crowberry
USES	Culinary, medicinal, household
PARTS USED	All mature parts are poisonous; very young shoots may be unsafe.
HEIGHT	4 to 10 feet, treelike, deciduous; thick, hollow, reddish smooth stems
FLOWER	White, sometimes purplish, small, slender terminal clusters; June to September; flattish, near-black, purple berries
LEAVES	5 to 12 inches long, oval, pointed, lance shaped
ROOT TYPE	Large, deep, fleshy
HABITAT	Hedgerows, edge of woods, roadsides, waste places
PROPAGATION	Seeds
CONTROL	Regular cultivation, digging up roots
CONTRAINDICATIONS	Unsafe, extremely poisonous

Pokeweed is a kind of vagabond among weeds, quite apt to turn up in unexpected places, even in gardens, such as near or among fencerow plantings along a property line, or around the periphery of a woodland clearing. Being of a handsome stature itself, not without a certain grace and charm, it has persuaded many a generous landowning soul to let it stay awhile. There is no doubt that the plant visually rewards such generosity, particularly in autumn, when its striking, large leaves turn color, its stalks assume a deep violet hue, and the loose clusters of berries become a deep purple.

However, beauty is as beauty does, and, on closer examination, pokeweed soon reveals one of its less appealing characteristics, that of emitting a most unpleasant odor. This kind of duality pervades the weed's entire existence. Not only does its odor diminish the enjoyment of its beauty but its several worthwhile, beneficial properties are tainted by the knowledge that they are also prepared to be dangerous. Published references tend to be ambivalent—some describe pokeweed as a medicinal or a sometime food plant, others give it short shrift as an outright poisonous plant.

Even the botanic name of pokeweed inspires a degree of perplexity. It refers neither to poison nor to medicine, yet the *phyto* part of the word is Greek for "vegetable or plant." *Lacca*, on the other hand, is the Latin adaptation of the Persian *lak*, or the Hindu *lakh*, both of which refer to a "purplish red pigment prepared from a resinous substance." In relation to pokeweed, this clearly alludes to the purplish red juice of its berries, and possibly, to the coloration of the plant itself. In French, the word came to be spelled *laque*, from which the English language, never slow to expand its vocabulary, fashioned *lacquer*. The specific name, *americana*, relates to the plant's origins, whereas *decandra* is a composite of two Greek words, to mean "ten-stamened," in reference to the weed's flowers.

Pokeweed is that relatively unusual entity—a truly indigenous American weed. It ranges from Maine to Florida, and westward to the Great Plains, even to Texas, parts of California, and Hawaii. It rarely, if ever, stands alone; instead, it enjoys the company of the bushes and trees bordering roadsides, or lounges about in neglected places; or else it rises above a tangle of tall grasses and other weeds in hedgerows and thickets, in clearings and waste ground. Although pokeweed is most contented in full sunlight, it is quite prepared to tolerate a modicum of shade. It is adaptable to various growing conditions but prefers rich loamy soils, except in coastal areas, where it is partial to life among the sand dunes.

Even when it is quite young, pokeweed resembles no other weed. I personally believe that it is a frustrated tree: Even the smallest first-year growth is shaped like one, and by the time a pokeweed root has been left undisturbed for a few years, the trunklike stalks it sends up may have a treelike diameter of as much as 4 inches. These stalks, varying in color from green to a deep magenta, and covered with large, fleshy, lance-shaped leaves, even branch and spread much like a tree or shrub. It is not at all uncommon for some of the leaves to become 10 inches long. Yet all illusions end in autumn, when pokeweed is revealed for what it is (a deciduous weed) and what it is not (a tree or shrub), when its leaves and branches wither and drop, and the "trunks" become dry and brittle hollow reeds, like cornstalks.

The American Indians who once inhabited, farmed, and hunted the eastern regions of the United States had known the weed long before the *Mayflower* landed at Plymouth Rock. They ate the *very* young shoots of the plant in spring, and they dug up the roots for medicine. With the purplish-red juice from the flattened round berries, they decorated their clothing and colored some of the crafts and adornments they fashioned from wood, bark, animal skins, and grains. Sometimes they even decorated their own skin with it. And the tribes living along the Connecticut River stained their splintwood baskets a rich, deep blue with the juice. The Delaware tribe believed that ingesting small portions of poke root cured rheumatism, whereas the Iroquois applied the root externally to treat skin disorders.

Inevitably, the early settlers adopted many of these uses to suit their own needs and quickly recognized that pokeberry provides a fine natural ink of a remarkably durable quality. Proof of that is the fact that the ink can still be seen and read on many a document preserved under glass in numerous museums. Just as inevitably, of course, the early settlers learned quickly that poke is a deceptive plant, that it can harm as well as charm.

There is no doubt that pokeweed possesses many virtues, but it is always wise to bear in mind its vices. As beneficial as it can be in treating certain afflictions, it is capable of being dangerously toxic. An overdose of poke root or pokeberries taken internally can not only induce severe vomiting and convulsions but may even lead to death. Even collecting a large harvest of the roots without wearing gloves can cause its toxic substances to be absorbed by the skin. And the seeds are considered particularly dangerous if swallowed. The same is true of the mature greens of the plant, its root, and all its other parts.

From a medicinal standpoint, however, the thick, fleshy, bitter-tasting poke roots have long been regarded of considerable value in numerous complaints. Gathered in autumn, they are cut into pieces and dried for the pharmaceutical trade, as are the berries.

Phytolacca, made from the dried roots and berries, is an extract or tincture that dairy farmers used to be able to buy at the local drugstore, and they mixed it with lanolin in order to treat the caked and swollen udders of their cows. Some chicken farmers used to, perhaps still do, plop a piece of poke root into their poultry's drinking water, in the belief that this prevented disease. Like other feathered friends, chickens and other barnyard fowl are not averse to pokeberries, but too many of the latter may not only give the meat a rather unpleasant flavor but imbue it with a laxative quality.

Although the preparation of home remedies using pokeweed is strongly discouraged, extracts of the plant, in the form of capsules or a tincture, and in mixtures of medicinal herbal teas, are prepared under controlled conditions and are prescribed by homeopaths. In one or another of these forms, it has been used for catarrhal conditions and diseases of the respiratory tract, as a laxative, or to reduce arthritic pains and inflammations, as well as for laryngitis, tonsillitis, and mumps.

Claims that the root is, or may be, of value in the treatment or prevention of cancer have been debated for more than 150 years, but research into this question continues. However, there seems to be little doubt among herbalists that poke benefits a wide variety of skin diseases when it is prepared and applied as an ointment. In the form of a poultice, it is said to relieve hemorrhoids and to bring boils to a head.

For the external treatment of chronic arthritis, swellings, or sores, early American settlers are reported to have prepared an ointment they learned about from American Indians. In a base of lard, or other grease, and beeswax, they apparently boiled together pieces of roasted poke root and elder bark, combined with bittersweet and yellow parilla (Canada moonseed). A simpler preparation, recommended for chronic rheumatism in the early nineteenth century, entailed infusing pokeberries in brandy.

Even the mature poke leaves, prepared in a salve, have been used to treat sore eyes. Applied as a poultice, they are said to have soothed the pain and itch from insect bites. And Euell Gibbons tells how his mother used to crush three dozen pokeberries in 1 pint of boiling water, and gave him and his siblings 1 tablespoon each of the cooled, strained infusion "to 'purify the blood' whenever we had boils or pimples."

Although it has been almost never cultivated in the United States, pokeweed was introduced into Europe in the late eighteenth century. There, especially in France and northern Italy, the succulent, *very* young shoots soon gained a reputation as a tender spring green at least equal to, if not better than, asparagus. In some instances, the plants were actively cultivated, and the spring shoots sold at city markets under the name of sprouts. The fresh sprouts at American markets were more an entrepreneurial than a strictly agrarian endeavor. American farmers who sold "shoots" had no need to cultivate pokeweed; they merely left it alone to develop and propagate at its leisure. Poke shoots are even included in at least one respected calorie book.

Today, young poke shoots or the leafy tips remain a desirable asparagus substitute in some foraging circles. Veteran wild-food gatherers caution, however, *never* to collect shoots that are tinged with red, and *never* to include a part of the root when severing the shoots. Speaking for myself, however, why should I run even the slightest risk to my safety and health when there are so many other, far safer, weeds for my veggies and herbal teas? Nevertheless, those who know about poke believe that, as long as the tender young shoots are collected before they are 6 inches high, are washed, trimmed, and thoroughly cooked in two changes of water, they are harmless. With the addition of "some salt and quite a lot of butter, margarine or bacon drippings," wrote Gibbons, "this is a delicious vegetable. It so closely resembles asparagus that some may be fooled." Gibbons also described pickling poke sprouts, preparing them as a leafy green vegetable, freezing them, and even cultivating them.

Still, even if pokeweed is banned from most dining tables and home-medicine repertories, there remains yet another option. This is to use the fruits as a dye or an ink by expressing their juice. Depending on the mordant used, the bright purple berry juice can color wool in tones that range from pink and rich red to brown or rust. The "ink" could also be used to design cards or letterheads or recipe folders, or to decorate gifts and crafts—or even to copy family documents. Who knows, in centuries to come, they might well be showcased in a futuristic museum. Only a word of caution: Although the juice is said to be harmless, the seeds are highly toxic!

If pokeweed has by now emerged as a complex vagabond, variously intriguing, repellent, helpful, harmful, good, and bad, a certain satisfaction can be derived from knowing that there was also a time in its history when it was made to account for its misdeeds by being itself victimized. It happened in Europe, after pokeweed had been introduced there in 1770. Suddenly, this New World traveler found itself at the center of a quite sophisticated fraud. While earnest Europeans studied its medicinal and culinary properties, and tested its value as a dye plant, there were others, less altruistic, who quickly pegged this import as their dream come true, their once-in-a-lifetime chance.

These particular opportunists happened to be French and Portuguese, although others would have undoubtedly practiced the same deceit, had they thought of it first. We will never know whether or not they employed the whole fruit, including the poisonous seeds, but these charlatans added pokeberries to color their cheap wines so that these would fetch higher prices. And as it is known that birds become intoxicated from eating too many pokeberries, it is not irrational to suppose that these "nouveau" wines had a similar effect on human bibblers. In any case, pokeberry juice spoiled the taste of the wines, and so the entrepreneurs' road to riches is said to have been blocked by a law before long, which required the innocent dupe of the scam, the pokeweed, to be cut down annually, before its berries could develop, or be used to tipple an unsuspecting public.

Plantago major

Plantago major
(COMMON PLANTAIN)

VITAL STATISTICS

Perennial

COMMON NAMES	Common Plantain, Greater Plantain, Ripple Grass, Broad-Leaved or Roundleaf Plantain, Slan-Lus, Snakeweed, White Man's Foot, Cuckoo's Bread, Healing Blade, Dooryard Plantain, Bird Seed, Englishman's Foot, Rat's Tails, Waybread
USES	Culinary, medicinal
PARTS USED	Leaves, roots, seeds
HEIGHT	12 to 20 inches
FLOWER	Tiny, purplish-brownish green, or yellowish green on long, erect spikes; June to October
LEAVES	To 9 inches long, to 6 inches wide, stalked, bluntly ovate, growing in basal rosette, grooved, thickish
ROOT TYPE	Dense cluster of thin, straight, yellowish roots
HABITAT	Lawns, fields, roadsides, waste places; full or partial sun, any soil
PROPAGATION	Seeds
CONTROL	Regular weeding

Safety in Numbers must be their motto. There are more than 200 members of the plantain clan in the world—all of them, so it seems sometimes, in my lawn and flower beds. Luckily, most of us need concern ourselves with only three or four of the species, and of these, by far, the best known is *Plantago major*, the common plantain. You undoubtedly know this rascally weed quite well. It's the one that leaves your lawn looking like a moth-eaten cushion, whenever you weed it out in summer, because you forgot to do so in spring; the one with the broad, flat leaves and spiky flower stalk shaped like an incense stick bearing an overdose of seeds. Or yours might be that other one, *P. lanceolata* (ribwort, lamb's tongue, snake plantain, jackstraw), which festoons a lawn with armies of lance-shaped leaves. These usually guard erect stalks in their midst, with far too many seeds in flower heads shaped like pointillist helmets.

Apart from their invading your lawns, plantains are usually found along roadsides, in fields, and in waste places. All of the species possess approximately the same properties and share similar characteristics, and all of them multiply at a formidable rate. *P. maritima* (sea plantain, sheep's herb) is prevalent in coastal regions, whereas *P. media* (hoary plantain, fireweed) turns up quite often in orchards.

Farfetched as it sounds at first, *P. media* is an instance of one plant being capable of healing another—fruit trees, specifically. Earlier this century, it was discovered that the leaves of this plantain, when rubbed on the part of a fruit tree affected by blight, effected an immediate cure. I tried it myself—skeptically, I admit—by tying a kind of poultice of bruised plantain leaves to an ailing cherry tree in my garden. It took several days, but the wound healed.

Leaves, roots, and seeds of plantain, fresh or dried, have been used for centuries in the treatment of such varied ailments as poison wounds, asthma, dysentery, earache, and kidney disorders. Rich in potassium salts, plantain is regarded as one of the finest remedies for cuts, abrasions, infections, ulcerations, and chronic problems of the skin.

Although its exact origins are uncertain, plantain seems always to have straddled the fence between history and legend. The Romans called it *planta*, "the sole of man's foot," because it followed Roman legions wherever they trudged to conquer the known world. Northern Europe, of course, was one such region. Briefly, when I was in Denmark some years ago, I felt an incredible nearness to those times when I learned that pollen analyses made of bog materials from ancient burial sites in Jutland had proved beyond doubt that plantain was introduced into that region during the so-called Roman Iron Age.

St. John the Baptist included plantain among his healing herbs; Pliny the Elder extolled its remarkable virtues in Rome, while Dioscorides recommended it in Athens for treating ulcers of the leg. And for early Christians, it came to symbolize the path to Christ. Inevitably, as the centuries passed, the wanderlust of plantains continued to be tweaked whenever colonists went in search of new territories. More often than not, native populations named it Englishman's foot or white man's foot.

Wherever plantain went, the word about its almost unbelievable healing powers was close behind. So it was bound to find its way into legends and literature. Anglo-Saxons called it the "mother of herbs," intoning the following magical verse whenever they applied plantain to a wound: "Carts creaked over you, queens rode over you, brides bridled over you, bulls breathed over you: all these you withstood, so may you withstand poison and infection." Through the *Lacnunga*, the most ancient source of Anglo-Saxon medicine, the recipe for a "salve for flying venom," using waybread, has come down to us:

> Take a handful of hammer wort [pellitory] and a handful of maythe [cam-omile] and a handful of *waybroad* and roots of water dock, seek those which will float, and one eggshell full of clean honey, then take clean butter, let him who will help to work up the salve, melt it thrice: let one sing a mass over the worts, before they are put together and the salve is wrought up.

Pliny claimed to have it "on high authority" that if plantain is "put into a pot where many pieces of flesh are boiling, it will sodden them together." In his *Colloquia*, the Renaissance humanist Erasmus recounted the tale of a toad that, to avert the poisoning effects of a spider bite, ate a plantain leaf. Shakespeare wrote of plantain's reliable healing power in no fewer than four of his plays, as did Chaucer in one of his works a century before the Bard. In France, to ensure their immunity before attacking a viper, weasels were said to roll in clumps of plantain. In the United States even today, some American Indians still wear plantain as a charm against snakebites.

Although I find the very young leaves of plantain pleasantly refreshing in spring salads, or steamed as a vegetable, or added to soups and stews, I would recommend eating them sparingly: They are mildly laxative!

Essentially, however, plantains are considered medicinal plants that are somewhat astringent, refrigerant, and diuretic, as well as vulnerary. The leaves, whether bruised, crushed, or chopped, have long been used by some Europeans as an emergency dressing for light injuries, and to stop bleeding. Washed in boiled warm water and crushed, plantain leaves are applied as poultices to badly healing wounds and sores. Slightly chewed, sprinkled with cayenne pepper, and applied as a poultice, a plantain leaf is said to have the extraordinary power to draw out any foreign object embedded in the body. If plantains are nearby—and they usually are—I always apply their crushed leaves to insect bites and stings, or to the portion of skin that has just experienced an unfortunate brush with stinging nettles or poison-ivy vines; plantain leaves also, I have discovered, cool the pain from burns and scalds. "One might say," wrote the wise Father Kneipp, "that the plantain closes the gaping wound with a seam of gold thread; for, just as gold will not admit of rust, so the plantain will not admit of rotting and gangrenous flesh."

In England, a favorite ointment for burns and raw surfaces in the late nineteenth century consisted of "boy's love" (southernwood), plantain leaves, black currant leaves, elder buds, angelica, and parsley, finely chopped and pounded, and simmered with clarified butter.

Also in England, plantain was at one time a standby emergency treatment for amoebic and bacillary dysentery. Three to four ounces of the root and leaves were brought to a boil in a pint of water, simmered for 4 to 5 minutes, covered, and infused for another 10 minutes. This tea was administered as often as the patient desired, in doses of 1 teacupful. Today, too, an infusion of the leaves (1 ounce herb to 1 pint water), drunk in doses of 1 cupful 3 or 4 times daily, can be prepared for individuals suffering or convalescing from bronchitis, whooping cough, ulcers, diarrhea, hemorrhoids, and excessive menstrual discharge. It is reputed to neutralize stomach acids and is considered beneficial in kidney disorders. Not least, the powdered, dried leaves taken in a drink were said to destroy worms. Applied externally, an infusion of plantain leaves is recommended for its soothing effect in cases of shingles and ringworm. And, while plantain does not rank among cosmetic herbs, an infusion of the leaf can be used as a most refreshing and stimulating skin lotion.

Although he particularly referred to *P. lanceolata*, because this is the most common plantain in his part of Europe, Father Kneipp could not have made a more impassioned appeal for plantain. If only people would gather these medicinal leaves, he said, and drink their expressed juice regularly, "numberless interior complaints, which shoot up like poisonous mushrooms out of the impure blood and the impure juices, would not arise. Those are wounds which, truly, do not bleed, but which are in many ways more dangerous than bloody ones." Nearly 400 years before Kneipp, Culpeper asserted, "All Plantains are good wound herbs to heal fresh or old wounds, or sores, either inward or outward." Among the plant's many virtues, he also claimed that "the juice, clarified and drank for days together. . . . staunches the too free bleeding of wounds." He also agreed with John Gerard, who had written long before, that "the juice dropped in the eies cooles the heate and inflammation thereof." And as recently as in my childhood, I recall a neighboring farmer's wife, in Bavaria, treating her children's earaches by pouring a few drops of the juice in their ears at bedtime. Not least, plantain juice has always been—and still is—deemed an excellent diuretic.

Because of the mucilage they contain, plantain seeds are often recommended as a gentle purgative in cases of constipation. Most effective of them all are the plantain seeds available from herbalists under the name psyllium. To prepare them, 1 teaspoon of seeds in 4 ounces of cold water is allowed to swell for 1 or 2 hours, then is drunk just before the last meal of the day. It is a somewhat thickish drink, its consistency not unlike bisque. Or, for their ascribed demulcent action on hemorrhoids, the seeds are often boiled in milk and drunk at bedtime. The seeds, first soaked in a little water, can also be applied on poultices. A syrup made of plantain seeds (1 ounce seeds boiled in 1½ pints water until reduced to 1 pint, then sweetened with honey) and taken in doses of 1 tablespoon 3 or 4 times daily is said to ease the pain of thrush, a common mouth infection in small children.

Plantain roots are an old-time cure for toothaches. Fresh, the roots used to be chewed, dried and powdered, and placed in a hollow tooth, the roots were reputed to be an admirable painkiller.

In the United States, the American Indians perhaps made the most extensive use of the plant. Early on it earned among them a name that is equivalent to "snakeweed"—the Omaha called it *ginebiwuck,* and the Chippewa knew it as *omikikibug.* From among the latter, Frances Densmore recorded an incident in which a woman was bitten by a venomous snake while berrying. By the time her husband had applied a crude tourniquet above the wound and found a plantain, the woman was in mortal danger. Yet, moments later, she was safe, after her husband had hurriedly cut little gashes into her badly swollen arm, and applied the crushed plantain leaves directly to them.

The Chippewa used plantain leaves to draw out splinters from inflamed skin, and as vulnerary poultices. Above all, they favored the fresh leaves, spreading the surface of these with bear grease before applying them, and renewing the poultices when the leaves became dry or too heated. Sometimes they replaced the bear grease with finely chopped fresh roots, or else applied the chopped roots directly to the wound. For winter use, they greased fresh leaves and tightly wrapped stacks of them in leather. The Iroquois, too, used the fresh leaves to treat wounds, as well as coughs, colds, and bronchitis. The Shoshone applied poultices made from the entire plant to battle bruises, while the Meskawaki treated fevers with a tea made from the root.

Not surprisingly, among the American Indian tribes, just as it has among many peoples around the world, the *planta* came to live in the dual world of fact and fantasy. Perhaps even now, a few of the tribal elders wear a charm against snakebites around their necks, in the form of a plantain leaf, fresh or dried, packed in a small leather pouch. After all, this charm has worked for 2,000 years, and, who knows, future generations may find it on the moon.

Portulaca oleracea

Portulaca oleracea
(P U R S L A N E)

VITAL STATISTICS

Annual

COMMON NAMES	Purslane, Pussley, Pursley, Pigweed
USES	Culinary, medicinal
PARTS USED	Whole herb
HEIGHT	To 6 inches
FLOWER	Yellow; spring, intermittently throughout summer
LEAVES	Spoon shaped, thick, fleshy, shiny dark green
ROOT TYPE	Small, fibrous
HABITAT	Fertile, sandy soil, sunny areas
PROPAGATION	Seed and rooting stems
CONTROL	Careful weeding

When in truly embarrassed financial straits, a reporter once suggested, one could even eat purslane—but, his article concluded, "don't invite me." That was written in the 1950s, in the days before nature was rediscovered, when some people thought that lamb chops grew on trees and when nature lovers were apt to be described as loonies.

The exact origins of purslane are uncertain; however, the weed is known to have been valued as a food by East Indians and Persians more than 2,000 years ago. Along its travels, it came to be respected not only as a vegetable and medicinal plant but also as an antimagic herb. In this latter capacity, purslane was believed to guard against evil spirits—provided it was strewn around a person's bed. In fact, if we are to trust an unattributed assurance of centuries past, purslane was even a sure cure for "blastings by lightening or planets and burning of gunpowder."

As the plant spread through most of the world, it became as native to parts of Europe and Asia as to the Americas and parts of Africa. Already long before it was introduced into the United States, purslane was considered a favored potherb in countries as different as China and Mexico. In many European lands, it enjoyed a respectable social position for hundreds of years, lovingly cultivated into several strains for their versatility as a nutritious garden vegetable. Although purslane never became indigenous there, it was just as enthusiastically raised in England, as is apparent from John Gerard's writings. "Raw Purslane," he opined four centuries ago, "is much used in sallads, with oil, salt and vinegar. It cools the blood and causes appetite." Nearly a century later, the English diarist John Evelyn added that especially the golden purslane (*Portulaca sativa*) is "generally entertained in all our sallets. Some eate of it cold, after it has been boiled, which Dr. Muffit would have in wine for nourishment."

Only in the United States has fame and favor eluded this hot-weather weed, although, inevitably, once it was introduced by early immigrants, purslane did what it has always done—it kept traveling. Nevertheless, popular sentiment seems to have largely agreed with nineteenth-century essayist Charles Dudley Warner, who described purslane as "a fat, ground-clinging, spreading, greasy thing, and the most propagatious plant I know." Today, it ranges freely between the Atlantic and Pacific Oceans, from Canada to Mexico, and it is difficult to imagine an area of human habitation in this country where purslane has not settled down. Yet, even when the inveterate forager Euell Gibbons proclaimed purslane as "India's gift to the world" in the early 1960s, the American public remained indifferent; nor has it changed its attitude in the decades since then.

Despite the fact that purslane prefers fertile, sandy soil, it is not at all averse to making its home in less hospitable, or different, conditions. Consequently, likely as not, it appears quite often in formal garden beds as a pesky weed. Allowed to remain there unchecked, this escaped garden plant will soon become a nuisance that is difficult to eradicate. One major reason is that purslane learned long ago to survive, and it does so through what might be called a no-fault propagation method—with you being the victim. I haven't yet decided whether this is purslane's way of punishing us for remaining stubbornly ignorant of its virtues or whether it is the weed's way of trying to win us over with nothing short of saintly patience. I confess that I first pondered this question when it dawned on me one day that purslane always grows more prolifically and succulently in my vegetable bed than anywhere else.

In any event, merely pulling purslane from the ground is not enough, especially when it is in flower, because the plant has the capability—with its dying breath, so to speak—to divert the food and water stored in its stems into the hurried completion of the seed production process. Just to keep us on our toes, purslane also relies heavily on its support troops. These are the portions of stem, however small, that our weeding might have broken off accidentally and left in the soil. Almost overnight, these portions make themselves known to us in the form of new emergent plants.

With such assertive characteristics, it would be easy to imagine purslane to be a plant of some stature. Instead, probably all of us have crushed it underfoot at one time or another. Depending on how happy it is in its immediate surroundings, a single purslane plant may attain a diameter of more than 1 foot, although it rarely reaches 6 inches in height. Its smooth, round, brittle stems are of a reddish color, radiating from the plant's center, forking and rooting repeatedly as they creep along the ground. Its tiny, yellow flowers, which appear solitary or clustered in the forkings of the stems and among the top leaves of the branches, bloom in early summer, opening only on sunny mornings. On the whole, purslane prefers full sun, although it manages quite adequately in some shade.

In spite of its spreading growth habit, unless it has overwhelmingly invaded a garden bed, purslane lacks the outstanding features that make us recognize and remember such other weeds as chicory (for its brilliant blue flower), ground ivy (for its balsamic fragrance), or stinging nettles (for their diabolical weapons). And yet, there is something immediately appealing about purslane, the kind of simple, direct, no-frills invitation that is pictured by a bowl of steaming hot soup on a cold winter's day, for example. And as a food is what purslane is known for most of all.

Its specific name, *oleracea*, from the Latin *holus* or *olus*, meaning "vegetables or potherbs," is the first clue that we are in the company of a savory morsel. All parts of it—leaves, flowers, buds, and stems—are edible, and can be served in a variety of ways, alone or mixed with other foods. Purslane can be used raw or cooked, frozen or pickled, or added to stews and soups. Even the unruly mob of tiny black seeds can be harnessed for productive KP duty.

As regards purslane, my own greatest pleasure comes from the double satisfaction of having a single task fulfill two goals: Every purslane plant I pull for the dinner table is a purslane I've weeded from my garden. And it's a free meal, a highly nutritious one that provides vitamins A and C as well as iron, phosphorus, and calcium. In fact, a leading botanist of the United States Department of Agriculture stated in the 1930s that, eaten as a potherb, purslane "is very palatable, still retaining, when cooked, a slight acid taste. It can be heartily recommended to those who have a liking for this kind of vegetable food."

If purslane is not a weed problem in your garden, but you happen to discover a few plants of it out of harm's way, then by regularly pinching off the leafy tips you can assure yourself of an ample supply, should you want it. Probably most often, purslane is used in spring salads, with or without other salad plants. I find that, dressed with oil and vinegar, the juicy mucilaginous leaves and stems add a mildly acid, piquant flavor. The leaves and stems can also be added to potato, grain, bean, or meat salads. When the weed is in bloom, the leaf tips make attractive decorations on cold platters, atop rice or mashed potatoes, as well as on plain vegetable dishes, as a change from parsley or chives.

Older shoots, steamed or boiled for a few minutes, are considered a fine substitute for plain spinach, lightly salted and peppered, and served with butter. Or, steamed and cooled, the green shoots can be prepared as a cooked salad. For a somewhat more elaborate vegetable dish, in a large skillet I sometimes sauté a chopped onion in a little butter or bacon fat, then add roughly 1 pound of purslane, season lightly with pepper, salt, and lemon juice, quickly stir, cover, and steam for about 5 minutes. I have also discovered that the mucilaginous quality of the plant—similar to that of okra, which some people might find objectionable in purslane served on its own—becomes a desirable binding element in vegetable or meat stews and soups.

Probably one of the simplest and most refreshingly satisfying summer soups can be made by stirring 1 cup each of purslane and sorrel tops in a little butter over medium-high heat for 2 to 3 minutes, then placing the greens in individual soup bowls or a tureen and pouring in heated, seasoned chicken or beef broth.

For a light creamed soup, I sauté 1 chopped onion in a little butter until translucent, then stir in roughly ½ pound each purslane and sorrel leaves and stems, cover, and cook for 5 to 8 minutes. Over medium heat, I stir in 2 table-spoons flour, until pale golden, beat in 2 cups chicken broth, then purée in a blender until smooth. Adding 2 more cups of broth (or 1 cup each broth and light cream, or milk), I season, reheat, and serve this nourishing spring soup.

Of course, if you already have a glut of other fresh produce from your garden, like me you may want to stockpile purslane for winter. To do so, I place cooked purslane meals in containers and freeze them until that inevitable dreary January night when it seems that winter will never end.

As summer advances the stems grow fatter; when the plants begin to go to seed, it is time to think of pickling purslane, ready for use with meats and cheeses. In fact, numerous European countries, particularly Holland, cultivate the weed for this purpose. Recipes vary, but here is one I like, because it is quick and easy. I loosely pack thoroughly washed and drained purslane stems into two clean pint jars and pour over them (leaving ¼ inch headspace) a mixture of ½ teaspoon each mustard and celery seeds, ¼ teaspoon peppercorns, 1 clove garlic, 1½ cups white vinegar, 1½ cups water, and 4 tablespoons salt that has been brought to a full, rolling boil for 1 minute. I store the sealed jars in the refrigerator.

Finally, since its earliest times, purslane has also donated yet another part of itself to humankind. In fact, this is something the American Indians of the Southwest discovered, too, once purslane crept into their lives. Among these tribes especially it was common practice not only to eat the plant raw but also to use its seeds in making breads and gruel. Their method for gathering the seeds remains much the same today. Once the flowers begin to fade, they gather quantities of the entire plant and let them dry on sheets of newspaper for several weeks. This is the time when purslane proves its ability to complete seed production even as the plant dies. When the seeds are ready, the dried plants are pounded to loosen and separate the seeds through a sifter. The final step is to grind this harvest of minuscule black seeds (in a hand or electric mill). Mixing the purslane meal with an equal portion of wheat flour results in the makings of what Gibbons describes as "dark but good 'buckwheat' cakes."

Although this last might seem as too much bother for perhaps no more than a single breakfast, purslane seeds were also highly recommended, by Nicholas Culpeper, for medicinal purposes. "The seed," he wrote, "is more effectual than the herb. . . . bruised and boiled in wine, and given to children, [it] expels the worms." Purslane was also said to soothe "heat of the liver" and "hot agues," and to be excellent for pains in the head "proceeding from heat, want of sleep, or the frenzy." The bruised plant "applied to the forehead and temples, allays excessive heat therein, that hinders rest and sleep; and applied to the eyes it takes away inflammation in them." There are also several claims, including by John Gerard, that purslane leaves eaten raw are beneficial when teeth are "set on edge with eating of sharpe and soure things." The distilled water of purslane, too, was said to ease toothache. And Culpeper wrote that the juice of the herb, mixed with sugar and honey, relieves dry coughs, shortness of breath, and immoderate thirst. Not least, together with oil of roses, the juice of purslane is said to be an effective treatment for sore mouths, swollen gums, and—dentists beware!—for fastening loose teeth.

Even if, in spite of its many virtues, purslane still never crosses your lips, I would like to think that you will be more kindly disposed toward this tenacious and resourceful little plant whenever you pull it from your garden in the future.

Rhus glabra

Rhus glabra
(SMOOTH SUMAC)

VITAL STATISTICS

Perennial Shrub or Tree

COMMON NAMES	Sumac, Smooth Sumac, Scarlet Sumac, Upland Sumac, Pennsylvania Sumac
USES	Culinary, medicinal, commercial
PARTS USED	Berries, bark, root
HEIGHT	4 to 15 feet, straggling, stout, hairless twigs; smooth, pale gray bark, reddish tint; inside pithy, with milky sap
FLOWER	Greenish red, small; in spring; tiny, dry, hairy, red fruit in dense upright, terminal clusters; June to October
LEAVES	Compound, 1 to 2 feet long; eleven to thirty-one leaflets, 4 to 5 inches long, lanceolate, toothed
ROOT TYPE	Rhizomic, suckering
HABITAT	Hedgerows, old fields, dry hillsides
PROPAGATION	Roots, seed
CONTROL	Vigorous cultivation

Smooth or scarlet sumac (often spelled "sumach") is not, strictly speaking, a weed, nor should it be confused with poison sumac (*Rhus vernix*), which is described below. Smooth sumac does not wantonly invade gardens everywhere, as do dandelions, for example, but it does tend to commandeer the borderlines of cultivated areas. From there, given the slightest opportunity, it will not hesitate to venture into a less than scrupulously patrolled garden. Fences are no obstacle to it. Its traveling roots will snake under them and advance several yards underground before sending up a young shoot, like a periscope checking to see if the coast is clear.

In this way, due to a period of neglect on my part and to opportunism on the part of sumac raiders, I found my berry patch besieged by them one year. Getting rid of them meant a good deal of work for me, digging up the roots and following them to their source. Yet, in another part of the garden, I have encouraged the growth of a small coppice of sumac.

In spring, when the earth bursts with a glory of colors, sumac does little to attract attention. It bides its time, capped by shiny, palmlike leaves, and modestly adorned with clusters of greenish red flowers that go virtually unnoticed. As the flowers fade, tiny hard, hairy drupes, or berries, develop into cone-shaped, red clusters at the tips of the branches, becoming increasingly visible. Soon, the leaves of the shrub change color, subtly at first, then turning more and more fiery, brilliant, every shade of red and golden yellow. And when the leaves have fallen, the clustered berries are fully exposed, like deep crimson plumes pinned to what looks like a tangle of antlers. They remain there through much of the winter, bright daubs of color against snow and sky and sepia shades—a feast of nourishment for wild birds to find.

Sumacs are a varied group of widely distributed shrubs indigenous throughout much of the United States. Besides smooth sumac, the best-known members east of the Rockies are staghorn sumac (*R. typhina;* formerly, *R. hirta*) and dwarf or winged sumac (*R. copallina*). Staghorn sumac is sometimes called velvet sumac, for its hairy stalks and twigs. In fact, smooth and staghorn sumacs are often virtually indistinguishable, because hybrids between them occur. *R. trilobata,* known as squaw-bush, because the American Indian women used its branches in basketry, is at home in the Rocky Mountain region. And indigenous to the Pacific coast are the elegant sumacs, *R. integrifolia,* sour berry, and *R. ovata,* sugar bush. These last two are quite different from the others, their growth more compact, their leaves leathery and entire, and their fruits borne in small, tight clusters.

All six of these sumacs grow wild, mostly in upland terrain, on hot, barren hillsides, in waste places, neglected fencerows, and old, abandoned fields. And occasionally, they are deliberately introduced into gardens, as specimen shrubs, or in groupings, or even as hedgerows. The fruits of all these plants are red and covered with a coating of fine and sticky acid hairs. The fruit of all six and the bark of the eastern species have long been prepared into tonic drinks and for other uses.

It is important to remember the preferred habitat of these particular sumacs, as well as the red color of their fruits. They are the good guys of the genus.

The all-too-infamous bad guys in the clan are poison sumac, *R. vernix,* as well as poison ivy and poison oak. Poison sumac is particularly perfidious, because, at first glance, it looks remarkably like the good guys—similar leaves, similarly colored bark, generally similar height. It also grows throughout most of the same regions as the three eastern sumac species described above. *However, beware!* The toxins of poison sumac are virulent. Contact with any part of the plant can cause even more unpleasant results than does poison ivy. Consequently, it is important to know the special features and habits of this enemy.

Poison sumac is a moisture-loving species that prefers to grow in wooded, swampy areas; its ripened berries are small and ivory white or grayish green, growing in loose, drooping or spreading clusters; the edges of its leaflets are untoothed and fewer in number, only seven to thirteen per leaf. And yet there is no denying that in its full summer growth poison sumac is often a remarkably lush and handsome plant, as long as it is admired from a distance!

The name *sumac* is based on the Arab word *summaq* and refers to any tree or shrub of the genus *Rhus,* and of several other genera. About fifty species of sumac exist in the world, variously in North America, in Asia, southern Europe, and in the Middle East. It is in the latter that I first encountered the red-fruited plant and learned of its use by Arabs as a substitute lemon seasoning. In England, one or another species is occasionally seen as the focal point of a suburban garden. Sometimes they even grow wild there.

Early fur trappers and plainsmen in the Ohio Valley and in the region of the Great Lakes referred to smooth sumac as *kinnikinnick*. It is an Algonquian word, meaning "mixture," because in American Indian practice a smoking "tobacco" consisted of the dried leaves and bark of certain plants, with or without the addition of an indigenous species of *Nicotiana*. The plants most commonly used were sumac leaves and the inner bark of a dogwood. But, as often happens when two cultures first learn to communicate with each other, the exact original meaning of *kinnikinnick* became blurred, and the pioneers adopted the word to mean the plants themselves—the sumac and red-osier dogwood or silky cornel, as well as others. Yet, before the Europeans made the act of smoking into a sociable habit, it had been, in American Indian cultures, strictly a religious rite, an offering of respect to elders, and a means of curing disease.

This had clearly changed by 1813, when the *Historical Dictionary* noted:

It has long been the practice among the natives of this continent, to substitute the sumach berry for tobacco, and the secret has been transmitted to Europe; in consequence of which it became so universally esteemed there by people of fashion and fortune, that large sums were offered to persons of mercantile professions, for this valuable but common production of nature. It has been preferred to the best manufactured Virginia tobacco. The method to be pursued in preparing the sumach berry to a proper state for smoking is to procure it in the month of November, expose it some time to the open air, spread it very thin on canvas, and then dry it in an oven, one-third heated. After having completed the progress of the cure thus far, spread it again on canvas, as before; there let it remain 22 hours, when it will be perfectly fit for use.

In 1887, a physician named Charles F. Millspaugh wrote that dried sumac berries were an article of trade in Canada, by the name of *sacacomi*, in the belief that when smoked, they served as an antidote to tobacco.

Today, sumac is probably no longer smoked, for health or any other reason.

However, another custom learned from the American Indians has endured. It is the making of "Indian lemonade," a wonderfully refreshing summer quaff that can be prepared from any of the six above-named *red*-berried sumac species. In fact, a common name for all three western species is lemonade-berry.

The main ingredient, the clusters of red berries or drupes, can be gathered from midsummer onward, while the tiny drupes are still densely covered with fine, sticky red hairs, giving the clusters a fluffy appearance. This stickiness is caused by the plant's acid properties, predominantly malic acid, the same as is present in apples, gooseberries, grapes, and rhubarb. In fact, it is these acids that give sumac berries their cooling (and reputedly diuretic) quality. However, because the active properties of both berries and bark are soluble in water, the ideal time to harvest comes during a prolonged dry spell, before heavy rains can wash away most of your lemonade.

Armed with a sharp knife or secateurs and a large basket—and a picnic lunch—harvesting sumac berries can be a pleasant Sunday afternoon spent outdoors. Once the "quarry" has been located, it is simply a matter of cutting as many ripened berry clusters as the basket can hold. I should have mentioned that bringing a book along is a good idea, because the entire harvesting procedure may take well under an hour, thus leaving plenty of time for reading. The moment the berry clusters are severed, a milky fluid seeps from the damaged bark, hardening almost at once into a gummy substance. This is considered harmless; in fact, it is said to have been formerly used in the treatment of urinary complaints.

There are two ways to prepare the lemonade. Either the clusters are first stripped of their drupes, then put in a large kettle, crock, or bowl, and covered with cold water; or, the entire clusters, covered with cold water, can be soaked in a large container. In either event, to extract most of the lemon-flavored acids, the berries should be stirred for 15 to 30 minutes. Next, it is essential to strain the liquid through at least a double layer of muslin, in order to remove all the fine hairs. The result is a most beautiful pink lemonade—and the know-how for an endless supply of it, at the cost of only the sugar or honey that are added to suit the individual taste.

The American Indians were so fond of this lemonade that they gathered large quantities of the berry clusters for their ample supply of the tart, cooling drink in summer, and dried them for a refreshing and healthful tea in winter. Like them, even modern trackers, foragers, hikers, and thirsty gardeners know that they have found welcome refreshment wherever sumac grows. All they have to do is to suck the acid coating from the ripe berries.

Dried and pounded, or finely ground in a blender, the berries are an excellent substitute for lemon flavoring. I sprinkle it on such foods as fish fillets, deviled eggs, egg or chicken salads, even fruits or light-colored puddings, as well as on green salads, of course. It looks like paprika and tastes like lemon—and usually wins new members for the sumac fan club.

An infusion of red sumac berries can not only provide a refreshing drink but is often used to ease feverish conditions. Such an infusion was formerly prescribed both for the treatment of diabetes and intestinal complaints. As a gargle, an infusion of the berries is used for infections of the mouth and throat; as a wash it can be applied to ringworm and ulcerations of the skin. Even a medicinal wine has been prepared from the berries.

The American Indians employed both the berries and the root bark for treating diarrhea and leukorrhea. They prepared a poultice of the crushed leaves for skin diseases and drank a decoction of the leaves to cure venereal disease. The Iroquois regarded the plant as an alterative (see Glossary).

The stem bark of *R. glabra* is cited as antiseptic and astringent, as well as tonic. A strong decoction made from it is also considered useful for diarrhea, and has been prescribed for gonorrhea, leukorrhea, and dysentery. Moreover, it is often employed as an antiseptic wash for skin disorders. Minor burns can be swabbed with a wash made from the bark of sumac boiled in milk. Taking 1 teaspoon of a decoction made from the bark is said to provide quick relief from throat irritations.

Apart from the pink "lemonade" it yields, and the smoking tobacco it used to provide, and apart from its services as one of nature's first-aid kits, sumac has also a respected history as a reliable agent in tanning leather and dyeing. Because of its high content of tannic acid, the plant has long been cultivated in the United States for the manufacture of an astringent extract that gives leather a yellow color. Could it be that sumac is the reason for all those armies of yellow work boots worn by men and women of every trade?

Yet, in spite of its prevalence and beneficence, sumac remains largely unappreciated. Just how much so was brought home to me only last winter by my dearest friend, Erika, a usually bottomless source of European herbal lore. She and her husband had come to see me, and together we took a picnic into a wintry woods studded with snowy crystals. We had just found a sunny spot to sit down, when, seeing a clump of scarlet sumac nearby, Erika exclaimed, "Be careful, I can't remember the name, but it's poisonous!" Like many others, at some point in her life she had obviously learned about *poison* sumac and had filed the name in her memory under "danger," unaware of the good-guy sumacs.

For me, however, besides all its practical uses, sumac—be it the smooth or staghorn, or one of their hybrids that grow in my area—serves one more purpose, which is purely frivolous. Indoors during the winter months, especially in dried arrangements, sumac never fails to attract admiration. Alone, or sparsely mixed with wild grasses, vines, berries, and seedpods, its deep crimson, plumed antlers pose with the nobility and grace of caribou in an arctic stillness.

Rhus radicans

Rhus radicans

(P O I S O N   I V Y)

V I T A L   S T A T I S T I C S

Perennial

COMMON NAMES	(*R. toxicodendron*) Poison Ivy, Poison Vine, Poison Creeper, Three-Leaved Ivy, Mercury, Picry; (*R. diversiloba*) Poison Oak
USES	Medicinal
PARTS USED	All parts are poisonous
HEIGHT	2 to 10 feet and more when climbing or trailing; hairy-stemmed vine, or smooth-stemmed shrub
FLOWER	Tiny, greenish, in small loose clusters; June; tiny, firm, white berries; August to November
LEAVES	Compound, three leaflets to each leaf; glossy or dull; slightly hairy or hairless; variously lobed edges; on long leaf stem
ROOT TYPE	Trailing, covered with brown fibers, short rootlets
HABITAT	Thickets, woods, hedgerows, waste places, borderlines of cultivated gardens
PROPAGATION	Rootstock and seeds
CONTROL	See text below
CONTRAINDICATIONS	Poisonous

It's a shrub. It's a vine. It's poison ivy, a member of the sumacs. It was formerly known by the specific name, *Rhus toxicodendron*, from the Greek words for "poisonous" and "plant" or "tree," although poison ivy was and still is unknown in Greece. It is an American native, and by far the most common of indigenous poisonous plants, often also referred to as poison oak. But poison oak is, in fact, a close relative, *Rhus diversiloba*, albeit an equally poisonous Pacific Coast species, although some botanists believe that both plants are variations of the same.

Probably no other plant in North America inspires such instant terror as does poison ivy at the mere mention of its name. Poison ivy may not be the most dangerous poison invented by Nature, but nobody who has had a close encounter with it will deny that it is the most fiendishly treacherous.

Of course, some people claim to be immune to its poison. I was one of them, fifteen years ago. Dressed in shorts, a short-sleeved shirt, thong sandals, and no gloves (!), I spent an entire weekend thinning out a densely overgrown hedge in my sister's garden while also wading through a lush ground cover of poison ivy. Cockily, I declined my sister's frantic invitation to the nearest hospital. I remained immune until, incredibly, and for no apparent or logical reason, I was forced to join the ranks of other poison ivy victims five years ago. Most recently, a string of blisters erupted on one of my hands, after I had literally done no more than to touch the *glove* with which I had pulled "dead" poison ivy off a tree trunk.

Protective clothing is clearly to be highly recommended—just in case—whenever one plans to launch an assault on poison ivy. That means long sleeves, fabric gloves, some form of head covering, slacks, socks, and closed shoes. This is warfare, you know, and the enemy is an old hand at it.

"Three-leaved ivy" comes closest to describing the plant, although it is, in fact, not three-leaved at all. Each leaf consists of three leaflets attached to a single leaf stem that emerges from the stalk of the plant. Unfortunately, recognition is often frustrated by the variations to be found in these leaflets. They may be smooth edged, lobed, or toothed; glossy or dull, dark or pale green; slightly hairy or hairless. Added to this, many people mistake the innocent Virginia creeper for this churlish villain; others confuse it with the woodland trillium, or even with the strawberry leaf.

Alas, only observation will eventually lead to instant recognition. Even so, I must confess that, seen with an unprejudiced eye, poison ivy is a rather beautiful plant, especially in autumn, when the leaves change color to every imaginable shade of red.

Poison ivy is usually described as preferring semishaded, moist locations, although I have discovered it countless times positively reveling in full sun and poor, sun-baked soil. What makes it trickiest to recognize by far is its variously climbing, trailing, and upright growth habits. It is usually found at the edge of woods, or in clearings, in dappled light; it also likes hedgerows—the more un-kempt, the better—and not infrequently, it skulks through the weeds along build-ing foundations.

The flowers, in June, are quite insignificant and appear in loose clusters. They are followed, from August until frost, by clusters of equally insignificant berries that are of interest only to certain birds. The latter, to their discredit, scatter the seeds in the course of nature's processes, so that new poison ivy plants may appear where you are sure they didn't exist before.

The vines are covered with coarse, dark brown fibers. Whether they trail under or on the ground, or climb up a tree, a building, a rock wall, or a fence, they do so with the help of countless rootlets that attach themselves to whatever surface is nearest, like barnacles to a ship's hull.

The poison of the plant consists of an acrid nonvolatile oil that is present in the entire plant, but is particularly prevalent in the leaflets and is secreted on their surfaces. Upon even the slightest contact, this irritating oil adheres to skin as well as to clothing. And now begins what can truly be described as torture, because poison ivy affects everybody differently, ranging from no reaction at all to un-bearable pain or even the need for hospitalization. In addition, a reaction or sensitivity to the poison may not manifest itself for several hours, or even for several days! If and when it does, it can be recognized by the eruption of painful, itching blisters that may not heal for as long as three weeks, sometimes more.

The blisters always appear linked on the skin's surface in a chain, or else clustered, much like the cells of a hornet's nest. Not scratching them takes almost superhuman self-control, whereas capitulating to the urge can result in the poison being spread to other parts of the body. I have found that the application of a reasonably hot, moist teabag (any of the commercially available brands) instantly soothes both pain and itch.

Should you ever become a poison ivy victim, do not for a moment think that you have, even after healing, fully paid the dues demanded for your unhappy brush with this vengeful devil of a plant. Not at all. If your *clothes* touched it, even though you might not wear them again for a year, you risk being poisoned again if your skin happens to come in contact with some of the poison. Therefore, I urge you to wash them all—especially the gloves—with naphtha soap and borax. If you used tools, wash them too.

Obviously, it is possible to avoid being poisoned by simply avoiding the plant, or by watching it grow from a safe distance, or by staying out of the woods entirely. A far more drastic measure is described by Euell Gibbons—for those who are truly rugged and reckless, I imagine. It is a regimen of gradual immunization, which requires the daily consumption of three young poison ivy leaflets, for a period of three weeks. As for me, I have too much respect for my lily-livered character.

Unlike most of the weeds in North America, the indigenous poison ivy crossed the Atlantic Ocean from west to east, to be introduced into Europe. It arrived in England in 1640, quite probably aboard one of the sailing ships returning from Plymouth Colony, and for 150 years, it appears to have kept a fairly low profile, both in England and on the Continent. Then, in 1798, a young Frenchman is said to have found himself suddenly, miraculously cured of a herpetic skin eruption he had had on his wrist for six years, after being accidentally poisoned by the imported ivy. Inspired by the cure, a physician at Valenciennes applied poison ivy in the treatment of other obstinate herpetic eruptions, as well as in that of palsy, apparently with some success. This soon led to the use of poison ivy in treating such other conditions as paralysis, acute rheumatism, and chronic eruptive diseases.

Although its medicinal action is described as irritant, rubefacient, stimulant, and narcotic, poison ivy won only very brief recognition in the *United States Pharmacopoeia*, and none at all in the British equivalent. In homeopathy, however, it did win acceptance for treating rheumatism, ringworm, and other skin disorders, as well as for treating incontinence of urine.

However, the level of my personal courage is such that at best I might conceivably allow myself to use the plant as an indelible ink—in desperate and dire circumstances only. After all, how does one extract the milky juice without first- or secondhand contact?

More sensibly, should you decide to rid your immediate surroundings of poison ivy, removing it by the roots is the most appropriate method, provided you use caution and common sense. Doing so in early winter is best, before heavy frost sets in, although the plant remains poisonous even then, only less virulently so. As the underground roots and vine may roam a considerable distance, it is essential to pull them up systematically, to ensure getting every scrap. Not doing so means the plant sends up new shoots from the remainder.

If you discover poison ivy in your garden, growing lush and cheeky, terminate its berry and seed formation by simply severing all the fibrous vine ropes at ground level. This will kill the aboveground plant, which you can then pry loose and remove in winter. The new shoots that arise from the severed vine roots next spring will show you exactly where to begin routing the underground network.

However, because this plant has the diabolically crafty habit of wending its way among tree roots, it is necessary for you, too, to fling aside all rules of common decency. A good shot of a systemic herbicide will eventually kill the weed. Hormone sprays and ammon-sulfamate may render it impotent in time. But the cheapest by far is a strong solution of borax and water, which destroys the weed. More than one application may be necessary.

Still, never trust poison ivy. Therefore, to make sure it is dead, cover it with plastic to rot, or hang it up to dry and die, and/or take it to the dump. *Never* add it to your compost. And whatever you do, *never, never* burn it. Breathing the fumes can land you in the hospital and may damage your lungs.

Poison ivy cures are almost without number. Washing the poisoned skin with gasoline is one; washing it with hot water and soap is another. Another surefire cure is said to be painting the afflicted skin with potassium permanganate (an oxidizer and disinfectant) in a 5 percent water solution. Although this results in turning the skin a hideous shade of brown, once the cure of the poisoning is effected, a weak solution of oxalic acid will get rid of the brown-skin stain. Still other recommendations are a dab of ammonia or even of bleach. Far less extreme, and just as available, are the sap of the aloe vera plant or the leaves of the plantain, which, when crushed, have proved most effectual in stopping the itch of ivy poisoning, according to the *New England Journal of Medicine*. Crushed or sliced garlic applied to the affected skin is also a superior treatment.

But the American Indians long ago discovered what still remains today by far the most popular prophylactic, as well as recommended home treatment, one which Nature itself quite often plants in the general vicinity of poison ivy. Its name is jewelweed, and it comes in two species—the orange-flowered *Impatiens biflora* and the yellow-flowered *Impatiens pallida*. Both plants are also called touch-me-not and snapweed, which refers to the habit of their slipper-shaped seedpods of bursting open at the slightest touch, and broadcasting the seeds a considerable distance.

Jewelweed is a tender, succulent plant that can grow to a height of 5 feet, preferring moist and semishaded locations. Its stems are almost translucent, pale green like its leaves, and gorged with a watery orange juice that seeps out freely when the stems are broken. This juice is believed by many to be the most reliably efficacious remedy for ivy poisoning. I once met a summer-camp teacher who claimed to have routinely given the children in her charge a daily cup of jewel-weed tea, which she made by infusing the plant. And at the end of every day, she apparently added a strong decoction of the weed to the children's baths. She claimed that during her tenure, the camp experienced not a single incident of ivy poisoning.

Oh yes, there *is* one good thing that can be said about poison ivy: Once it has been pulled up, it almost never returns.

Please note the *almost!*

Rumex acetosella

Rumex acetosella
(SHEEP SORREL)

VITAL STATISTICS

Perennial

COMMON NAMES	Sheep Sorrel, Field Sorrel, Red Sorrel; (*R. acetosa*) Common Sorrel, Garden Sorrel, Greensauce, Sour Sabs, Cuckoo's Meate; (*R. scutatus*) French Sorrel; (*R. hastatulus*) Wild Sorrel
USES	Culinary, medicinal, household
PARTS USED	Leaves, seeds
HEIGHT	12 inches high; other varieties to 2½ feet; slim stem rises from basal-leaf rosette
FLOWER	Reddish green or brown, clusters; June to September
LEAVES	Arrowhead shaped, smooth, yellowish bright green, turn red in autumn, variable breadths
ROOT TYPE	Perennial, yellow, rapidly spreading
HABITAT	Meadows, fields, roadsides, poor or sour garden soil
PROPAGATION	Seeds, roots, division
CONTROL	Weeding, soil improvement, liming
CONTRAINDICATIONS	Caution: Large doses of medicinal tea may cause severe kidney damage.

Sorrel is a weed of which I have clear childhood memories. It reminds me of mild, sunny days, whiling away lazy hours in the grass, staring at the sky, and idly chewing one of the sour leaves. It is a weed my sister and I picked often so that Mother could make one of our favorite soups. I also remember gathering it on special school "nature trips," for use in military field hospitals during World War II. Over the years since then, sorrel has often quenched my thirst, staved off occasional queasiness, and cooled and soothed the swelling from many a bee or other insect sting. Sorrel has also given me some giant headaches as a troublesome pest in my garden.

First of all, in spite of the fact that both are commonly called sorrel, there is absolutely no botanical kinship between the *Rumex* and *Oxalis* genera. Nature must have decided to tease us mortals by giving both of them similar properties and taste. Naturally, we humans promptly responded with a bit of good-natured humor of our own, by referring to almost *all* the species of both genera as "sorrel." The upshot of so much wit and subtlety is that some of us are left thoroughly confused. Luckily, Nature provided sufficient differences to make it possible for all of us to distinguish one genus from the other at a glance.

The *Rumex* species all have lanceolate leaves, whereas the *Oxalis* species are trifoliolate, composed of three heart-shaped leaflets folded along the center. Nevertheless, because of their similar constituents, both *Rumex* and *Oxalis* have been put to similar uses. This chapter, however, concerns itself only with the *Rumex* types of sorrel most commonly known in the United States. They are *R. acetosella*, *R. acetosa*, and *R. hastatulus*. Of these, the first is by far the most prevalent.

Make no mistake about it, sheep sorrel can be a serious nuisance, especially if it is allowed to dictate the terms of its own destiny. At this moment, there are two areas in my garden where this cheeky invader has made noticeable inroads. Of course, I am irked, all the more so because one of the areas is a large perennial border. On the other hand, the sorrel's presence in such numbers sends me a very clear message: The soil is sour; it needs nourishment, improvement, sweetening.

Sorrel spreads by means of its tiny seeds and via the myriad roots that often run deep underground, entwined among the roots of grass and other plants. The roots of a single sorrel rosette can travel several feet, sending up new rosettes at intervals; the roots of each of these, in turn, produce rosettes of their own, and so on. The resulting underground tangle can be a gardener's nightmare. In the meantime, naturally, the seeds are not idly lolling about either. In fact, having been launched from their capsules like an entire arsenal of miniature projectiles, they promptly set about producing rosettes of their own so that, given suitable conditions, sorrel can invade and occupy large areas in a single season.

Up to this point, you will notice that sorrel's habits are not altogether unusual in the realm of weeds. But what adds malice aforethought to its invasions is the fact that its roots break *very* easily. And each piece of broken root is capable of becoming a new plant. As I have too often discovered to my utter chagrin, attempting to extricate merely a single root may cause enough breaks in it to double the original number of plantlets!

With such unattractive traits, it is difficult to imagine that, for 1,000 years, sorrel has been the central feature in the commemoration of a great Irish victory over the Norse invaders. The battle took place in County Meath, northwest of Dublin, in A.D. 980. According to local tradition, when the sorrel turns crimson in summer, giving the meadows a reddish hue, the leaves of this lowly weed each year point to the graves of all the ancient patriots whose blood was spilled on the Hill of Tara.

It is equally surprising to learn that sorrel was already a food staple in Europe during the Roman Iron Age around A.D. 400. Irish peasants came to prepare it with fish and milk; Laplanders used the juice of the leaves as a substitute for rennet to curdle milk, and, in times of dire need, Scandinavians sometimes put sorrel into bread in order to benefit from the plant's small content of starch and mucilage. Early English peasants ate the leaves as a meat or green. To tenderize tough meat before cooking, they tied sorrel leaves around the cuts. Or they cut and mashed the leaves and mixed them with vinegar to prepare a dressing served with cold meat. Their more affluent countrymen added sugar to the dressing and called it "green sauce" to accompany meat and fish and fowl.

Over the centuries, sorrel climbed ever higher in social circles, until, during the reign of King Henry VIII, it was titillating the taste buds at Hampton Court.

Half a century after Henry's death in 1547, the herbalist John Gerard discussed it in his study of useful plants. "The juice thereof in summer time," he wrote, "is a profitable sauce in many meats and pleasant to the taste. . . . The leaves are eaten in a tart spinach." As a potherb prepared like spinach, it was an excellent accompaniment to veal, lamb, or sweetbreads. Cooked with turnips or parsnips, sorrel added piquancy to the otherwise bland flavor of these vegetables. Simply steamed, the tart flavor of the leaves counteracted the richness of goose or pork, and provided a pleasant alternative to applesauce.

By the eighteenth century, sorrel was a popular source of inspiration among London's literary, as well as gastronomic, circles. The most lavish praise came in 1720 from John Evelyn, who was convinced that sorrel imparted "so grateful a quickness to the salad that it should never be left out." Among its many attributes, Evelyn claimed, "Sorrel sharpens the appetite, assuages heat, cools the liver and strengthens the heart." In fact, he concluded, "Together with salt, it gives both the name and the relish to sallets from the sapidity, which renders not plants and herbs only, but men themselves pleasant and agreeable."

Poets, too, were charmed by sorrel. John Keats wrote, in his *Endymion*:

Cresses that grow where no man may them see,
And sorrel, untorn by the dew-claw'd stag;
Pipes will I fashion of the syrinx flag;

More than a century later, Algernon Charles Swinburne mused:

Simplest growth of Meadow-sweet or Sorrel
Such as the summer-sleepy Dryads weave,

while directness itself was *The Salad*, from William Cowper's translation of "Virgil":

There flourish'd starwort and the branching beet
The sorrel acid and the mallow sweet.

Alas, the popular garden sorrel gradually lost favor in England once the larger, more succulent French variety was introduced sometime after the death of

Henry VIII. However, it seems this merely signaled greater culinary inventiveness.

In France, sorrel was added to fricassees and ragouts; it formed the basic flavoring of a delicate consommé or broth. Even now, especially in rural parts of the country, it remains an essential ingredient of a *soupe aux herbes*, which was formerly offered by some of the finest restaurants in Paris. In fact, in *Modern French Culinary Art*, published in 1966, well-known French chef and cookery teacher Henri-Paul Pellaprat includes, among several sorrel recipes, the delicately delicious consommé chiffonade.

In America today, although they are not standard fare at restaurants, numerous wholesome dishes made of sorrel can be prepared at home at little cost. I include my mother's recipe for sorrel soup, because it has always remained one of my favorites. It is simple to prepare, as well as nourishing.

However, there is a caution. The sharp taste of this herb derives not only from its vitamin C content, but also from its content of binoxalate of potash, or oxalic acid. This same salt gives our common garden rhubarb its acid taste. Oxalic acid can be harmful if taken in large doses; therefore, some herbalists recommend that it is best avoided by the very old, the very young, and by people suffering from rheumatism, kidney ailments, and gout. On the other hand, the oxalic acid can be reduced by always parboiling the herb for two or three minutes and straining before cooking.

To prepare her sorrel soup, Mother first poured boiling water over 1 pound washed and finely chopped sorrel leaves, letting these steep for about 3 minutes, then straining them, setting them aside, and discarding the water. Over medium heat, she melted 4 ounces butter or margarine in a medium-size saucepan, in which she sautéed 2 finely chopped medium onions. Next, she stirred in 4 tablespoons flour until golden brown. With a wire whisk, she vigorously blended in 3 pints heated chicken broth (or water), until the mixture was smooth, then added the sorrel. She seasoned the soup with 1 tablespoon sugar, ½ teaspoon grated nutmeg, 3 tablespoons lemon juice, plus pepper and salt to taste, and simmered it for about 30 minutes, stirring occasionally. Sometimes she served it plain; at other times, she stirred in 1 cup of yogurt.

Sorrel can also be prepared as a creamed vegetable. Except for first steeping the leaves in boiling water for 3 minutes, this can be prepared exactly like creamed spinach. For a slight variation in flavor, a peeled, cored, and grated apple can be stirred into the cream sauce, seasoned with a dash of nutmeg. For yet another slight variation, the creamed vegetable can be served in pastry shells. I also give scrambled eggs or omelets a tangy new flavor by adding a few chopped, cooked sorrel leaves minus leaf stalks.

A refreshing drink can be made by extracting the juice from 8 ounces of sorrel leaves and blending it with 12 ounces buttermilk, sweetened with honey. And nothing could be simpler than a modern adaptation of "green sauce" with a handful of stalk-free sorrel leaves. In a blender or processor, the leaves are chopped and mixed with 4 ounces yogurt, 2 ounces sour cream, 2 ounces heavy cream, 2 teaspoons lemon juice, pepper and salt, and a pinch of sugar, then chilled. Garnished with parsley, the sauce can be served as a dressing with hot or cold fish, chicken, veal, eggs, or green salad.

As a medicinal plant, sorrel has always been a relatively modest herb. Nevertheless, it has been used for centuries as a diuretic, diaphoretic, and refrigerant. The juice extracted from the whole plant has also been recommended for urinary and kidney ailments. The plant is also considered mildly antiseptic and laxative. And popular opinion in times gone by seems to have held that it cleared at least one bird's vocal cords—hence the name cuckoo's meate.

Culpeper wrote of sorrel leaves being "very cooling, allaying thirst, and repressing the bile; good in fevers." He considered them of particular value against scurvy, if eaten in spring salads. Not least, he claimed, "both roots and seeds, as well as the herb, are held powerful to resist the poison of the scorpion."

Traditionally, too, sorrel tea made from an infusion of the leaves was given as a cooling drink against fevers, and to stimulate the appetite. In folk medicine, poultices made of cooked sorrel reduced to pulp are applied to boils, to bring them to a head. Similarly, either fresh crushed leaves, or a decoction of them used on compresses or as a lotion, are used to soothe badly healing sores and wounds, or minor skin afflictions.

The juice extracted from the leaves and mixed with a little vinegar or else vinegar added to a strong decoction of the whole herb is often gargled for a sore throat or used as a mouthwash. Both the root and the seeds were formerly used to stop hemorrhages, and the flowers decocted in wine were said to cure jaundice.

Sorrel is even ready to do KP duties. That is, the stalks I discarded when making the omelet can be chopped, added to a little water, shaken vigorously, and then used to clean out bottles or jars that have contained oil. The procedure may have to be repeated. Before discarding the water in which sorrel has been par-boiled, a cleaning cloth can be rinsed in it to wipe wicker furniture squeaky clean. Ink stains, too, are said to disappear if they are first rubbed with freshly picked sorrel, then washed in soap and rinsed. But just in case the ink stain ends up being replaced by a chlorophyll stain from the plant, I myself would first try this method on something old.

Still, taken all together, these varied uses to which sorrel can be put surely indicate that it is truly repentant, trying to atone for all the aggravation it is to gardeners.

Rumex crispus

Rumex crispus
(C U R L E D   D O C K)

Perennial

COMMON NAMES	Curled Dock, Dock, Yellow Dock, Sour Dock, Narrow-Leaved Dock, Bitter Dock, Out-Sting
USES	Culinary, medicinal
PARTS USED	Roots, leaves, seeds
HEIGHT	2 to 4 feet tall, rigid, knotted, brownish green stalks
FLOWER	Many, small, greenish yellow or reddish, clustered, atop long stalks, June to September; seeds in three-winged, one-fruit valves
LEAVES	6 to 12 inches long, narrow, lance shaped, with curled edges, large at base; smaller, fewer toward top
ROOT TYPE	8 to 12 inches, parsnip-shaped taproots, rarely forked, brown exterior, yellowish white interior
HABITAT	Rich soil, meadows, woodland clearings, waste places, roadsides, ditches, stream banks
PROPAGATION	Seeds, severed root
CONTROL	Digging up entire root before plant blooms
CONTRAINDICATIONS	Caution: Leaves may cause dermatitis.

If they ponder about it for a moment, even people who know neither the name nor anything else about curled dock will realize that they are familiar with it, because of its tall seed stalks in autumn. Covered with myriad tiny, brown, heart-shaped, winged fruit, these rise like spires above surrounding weeds or grass, beside highways, in waste places and pastures, and even in cities. Children delight in stripping the seedpods and scattering them, and flower arrangers gather the stalks for their ornamental value.

About thirty species of docks are widely distributed throughout North America, some of them introduced from Europe. Curled dock, bitter dock, and patience dock are probably the three that are most often encountered. All docks belong to the genus *Rumex*, and all of them possess similar properties. Although there are variants among them in leaf shape and size, in stem height and color, and in flower and fruit, their uses are said to overlap. Their mutual common name, *dock*, has a history almost as long as that of the plant's use itself, being derived from the Anglo-Saxon word *docce*.

Before they were assigned to the genus *Rumex*, docks used to be members of the genus *Lapathum*, a name probably taken from the Greek *lapassein*, "to empty," in allusion to the mildly purgative and blood-purifying properties attributed to the plants, for which they continue to be valued. All docks are also related to garden rhubarb, which they resemble to some extent, and with which they share a number of constituents, including tannin.

Curled dock is easily recognized by its long, narrow leaves, whose edges are "crisped" or "curled," which is the meaning of the plant's specific name *crispus*. What suggests that this dock may be more prevalent than the other species, and perhaps better known, is the fact that it has earned the attention of the United States Department of Agriculture, which has described it as a "troublesome weed." Without doubt, this is a sobriquet with which many a farmer or gardener heartily agrees. The reason is simple: The depth to which dock often sinks its taproot is proof enough of its well-developed survival instinct and its intention not to be easily dislodged.

Although docks prefer to plunge their roots—some of them 2 or 3 feet long—in rich soil, their presence is a useful indicator that the soil is sour and the drainage poor; the more docks congregate, the greater is the acidity. The farmer's or gardener's response to this may be twofold: reducing acidity by means of deep cultivation, soil aeration, and the addition of organic matter and lime; or neutralizing or altogether removing the roots.

Docks are spread by seeds, but because the rootstocks survive, even if they have been severed, these manufacture new plants. One remedial approach, therefore, is the frequent cutting of aboveground growth in order to prevent seed formation. Some say that this method eventually also starves the roots. Another approach, effective in small areas, is to pull up by hand, or with the help of a trowel or spade, each young dock before it can produce seed or develop large roots. However, it is essential to remove the *entire* root: Even a leftover piece is capable of sprouting not just one, but several new docklets. A third method of eradicating this weed, especially when it is old and well established, is to lop off the top 2 inches of the root and cover the fresh wound with about ½ cup of salt. The root will die.

All of this extermination talk, however, proves nothing so much as that the *real* trouble with "troublesome" weeds rests not with weeds but with us. Again and again, their stubborn persistence confounds and offends our sense of order; as a result, our own stubbornness often blinds us to the clues that weeds keep putting before us: There is more to them than meets the eye and we could benefit from a little control of our destructive instincts.

Its occasional nuisance notwithstanding, dock is really a good guy; in fact, curled dock's medicinal properties earned it a place in the *United States Pharmacopoeia* for several decades in the nineteenth century. Although it was removed in the 1890s, not only has it remained a much valued plant in modern herbalist medicine, available even in the form of gelatin capsules, but large quantities of the root are reported to be actually imported annually.

However, perhaps the weed's most continuous use has been as a mild purgative, a blood-purifying tonic. Referring to it as "Bloudwort," John Gerard considered it "of pot-herbs the chiefe or principall, having the propertie of the bastard

Rubarb, but of lesse force in his purging qualitie." Just as ancient herbalists knew this, so do modern herbalists, too.

Dock's roots are most often employed, fresh or dried, and prepared as an infusion and decoction, a syrup, an ointment, a powder, or a poultice. Or they are added in small amounts, finely chopped, to herbal teas, or even prepared as a healing wine. They also yield a yellow dye. The roots are generally harvested at the end of the summer and used fresh or dried. To dry, the thoroughly scrubbed fresh roots are cut crosswise into ½-inch rings, or split lengthwise in half. Spread on brown paper, the pieces are allowed to dry in the sun, in an airing cabinet, or in an oven on low heat, and stored in a tightly closed container in a dry location.

To prepare an ointment for external applications to skin disorders, wounds, or piles, herbalists used to boil the fresh or dried root pieces in vinegar, until the fibers were softened. They would then mash and mix the pulp in lard and store it in a tightly sealed container. In a similar preparation today, petroleum jelly can be substituted for lard, and sulphur may be added to the ointment.

Externally, too, a decoction of dock root is considered useful when applied as a wash or lotion, or used on compresses. Pulped, cooked dock roots have a long tradition as poultices directly applied to infected wounds and sores. This last is a treatment said to have been common among some American Indian tribes, notably the Navaho and Blackfoot, who used mashed roots of dock both for human wounds and swellings and to heal the saddle sores of their horses. An infusion of 1 teaspoon dried dock root steeped for 30 minutes in 1 cup of boiling water, and cooled, is still reported to be used in the Appalachian region to treat hives and ringworm.

By far the most drastic treatment for skin afflictions, however, appears to be one offered in the early seventeenth century by Gervase Markham, a noted authority on husbandry and farriery, who reputedly wrote that applications of wood sorrel "lapped in red Dock leafe and roasted in hot cinders, will eat away the dead flesh in a sore."

Judging from the numerous regional versions of the same English country rhyme, there must be truth to the folk belief that the juice of crushed dock leaves is an antidote to nettle stings.

Nettle out, dock in,

Dock remove the nettle sting

says one such rhyme. Certainly, the astringent property of the weed is commonly used to soothe the burning pain. Traditionally, it does not matter which particular version of the rhyme you recite, as long as you do so *slowly*, and while you are rubbing the dock on the afflicted area.

Taken internally, whether as an infusion, a decoction, or a syrup, dock's action is described as mildly tonic, diuretic, and laxative. It is mostly prescribed for diseases of the blood and for chronic skin ailments, general debility, and anemia. It is also given to stimulate digestion and to improve the flow of bile. It is said to nourish the liver and spleen, to aid in the cure of jaundice, and to be beneficial for arthritis and for lymphatic problems. It has been used to relieve a congested liver, a dry, tickling cough, and glandular swellings. It is also said to cure headaches and ringing noises in the head, to soothe itching ears, toothache, and inflamed eyes and eyelids, to stop nosebleeds, and to ease flatulence. The root of curled dock has even been considered of some value in restraining the advance of cancer in the human system. Today's homeopaths prescribe dock for inflammations of the larynx and trachea, and also for persistent dry coughs.

An infusion is made by pouring 1 pint boiling water over 1 teaspoon powdered root, letting it steep for 5 minutes, and straining it through muslin. For a mild decoction, 1 ounce of fresh root is boiled in 1 quart water for 5 to 10 minutes, steeped for an equal length of time, and strained. One small wineglassful morning and evening is usually suggested. For a syrup, 8 ounces fresh or dried root pieces are brought to a boil, then crushed in 1 pint water; 1 pound sugar or honey is stirred in, until dissolved. The syrup is gently boiled until it thickens slightly, then strained through muslin, and bottled. It is usually given in doses of 1 teaspoonful once or twice daily.

William Smith offers a recipe specifically for babies suffering from thrush. "A decoction," he writes, "from one ounce of Yellow Dock root and a quarter-of-an-ounce of Gingerroot (both cut or bruised), using just over one pint of water should be simmered for fifteen minutes and then poured over one ounce of Raspberry leaves whilst still boiling. Strain this liquid and divide into two parts, one part to be used as a cool mouth wash." What Smith fails to explain is how a baby young enough to have thrush can get the hang of gargling or rinsing its mouth.

Gerard enthusiastically recommended a brew that included a mixture of dock and other herbs and spices steeped in 4 gallons of ale. He further recommended taking this "as your ordinary drink for three weeks together at the least," adding that it purifies the blood "and makes yong wenches look faire and cherry-like." Were Gerard alive today, he would surely allow that such a brew might have similar effects on young lads!

To accommodate modern preferences for speed and efficiency, I include the following recipe for dock wine, which is not only reputed to encourage good health but must surely rank as among some of the less common ways to start a day. The recipe is French, its purpose is supposed to be strictly medicinal (of course). In a covered, nonmetal saucepan 180 grams dock root, 6 grams licorice, 3 grams juniper berries, 120 grams sugar, and 2 liters good red wine are macerated together for 2 hours. Still covered, they are brought to a gentle boil until reduced by one-third, then strained, and bottled. Of this dock wine, says Jean Palaiseul, "drink 90 grammes each morning on an empty stomach for fifteen consecutive days." That is just over 3 ounces each time! *Bonne chance,* say I.

As the Iroquois Nations of North America had already known long before the white man came, dock can also be used as a food. Even Culpeper reported that "all Docks being boiled with meat, make it boil the sooner; besides, Bloodwort is exceeding strengthening to the liver, and procures good blood, being as wholesome a pot herb as any growing in the garden." However, lodging his objection, he added, "yet such is the nicety of our times, forsooth, that women will not put it [dock] into a pot, because it makes the pottage black; pride and ignorance (a couple of monsters in the creation) preferring nicety before health." Tut, tut.

In spite of such vicissitudes, dock's history is such that the weed has even found its way into a museum. The Danish city of Silkeborg is not very large, and neither is its museum, which contains an exhibit that has left an indelible impression on me. It is known as "Tollund Man." Curiosity had taken me there to see the preserved head of a man who had lived and died 2,000 years ago. He had been found in a peat bog in 1950, and the skin of his face, though darkened by peat, had also been preserved by it. Looking as though he were merely asleep, and wearing a simple, fitted animal skin cap, with tufts of his hair showing above the forehead, Tollund Man's is one of the noblest, most beautiful, most peaceful faces I have ever seen. What fascinated me most of all was the stubble of beard on his chin and upper lip. He looked so alive that I fully expected to see him breathe, or to open his eyes.

And here I learned that Tollund Man's last meal, typical of the early Iron Age, had consisted of a gruel made from various cultivated and wild grains. Among these were barley, linseed, knotweed, bindweed, camomile—and dock.

Two thousand years later, dock is still valued as a nutritious food. For many people, winter only ends with a serving of an early spring salad made from the very young leaves of dock, either alone or mixed with other salad greens. As the very newest dock leaves appear in delicately colored hues, they can also be used to decorate such dishes as other salads, rice, potatoes, eggs, or meats. Once the stalks appear, however, the leaves become too tough and bitter to eat.

As a cooked potherb, the tender young dock greens are considered by many superior even to early spinach. Moreover, they are rich in iron and protein, as well as in vitamins A and C. But the greens' content of oxalic acid, which gives them a slightly bitter taste, can also exacerbate gout. The bitterness can be reduced by double-cooking the vegetable. That is, the first water (to cover) in which the leaves are boiled for 3 to 5 minutes is discarded, and the leaves are cooked another 6 to 10 minutes in a second boiling of lightly salted water. The greens can be served with lemon juice or vinegar and a little pepper, or with a béchamel sauce. For a more robust vegetable, a piece of bacon or salt pork can be cooked with the leaves. However, for those who are truly determined to purge their bodies of all the impurities accumulated through a long winter, a cooked dish of mixed dandelion, mustard, horseradish, and dock greens is regarded as a most effective solution.

With so much to recommend curled dock, why not entertain thoughts of sparing the next one you find in your garden? At the very least—you can never be sure—it may protect you against "elf sickness." The Anglo-Saxons believed it could.

Senecio vulgaris

Senecio vulgaris
(G R O U N D S E L)

VITAL STATISTICS

Annual

COMMON NAMES	Groundsel, Common Groundsel, Grundy Swallow, Ground Glutton, Simson, Sention, Birdseed, Ragwort
USES	Medicinal
PARTS USED	Whole herb
HEIGHT	6 to 12 inches, erect, green to purple stems
FLOWER	Small, cylindrical, yellow, clustered; March to October
LEAVES	Dull dark green, long, narrow, irregular, jagged, blunt toothed
ROOT TYPE	Small, white, fibrous
HABITAT	Fields, gardens
PROPAGATION	Seeds
CONTROL	Regular weeding
CONTRAINDICATIONS	Caution: To be used internally only under medical supervision.

G roundsel, a European native, may be only an annual, but it has efficiently utilized its existence by mastering the art of survival. As early as 2,000 years ago, it had devised its own expansion campaign, through which it has long since conquered much of the world, including most of North America. Unlike many other plants, which waited to be introduced around the globe by early settlers and colonists, groundsel took charge of its own travel plans. Again and again, it stowed among the food grains that accompanied all European migrations to foreign parts. Once arrived, this seemingly innocuous weed set about doing what it does best—it multiplied at a formidable rate. Because it is in flower and seed production throughout the growing season, a single groundsel plant is capable of bringing forth an estimated 1 million others within a single year.

In recognition of this prodigious spreading habit, the Anglo-Saxons named the plant *groundeswelge,* meaning "ground swallower." *Grundy swallow,* the name sometimes still heard in Scotland and northern England, is only a slight corruption of this. The names *simson* or *sention,* used mostly in eastern England, are probably corruptions of *Senecio,* the plant's botanic name, which is derived from the Latin *senex,* for "old man," in reference to the tufted white seed heads.

Throughout its travels and far-flung settlements, groundsel seems never to have been employed as food for humans; however, certain members of the animal kingdom have always been the weed's loyal fans—and consumers. Goats and pigs are especially fond of it; cows avoid it, if they can. To sheep it is nothing more than baaaa-humbug, and horses merely snort at the sight of it. But rabbits can be enticed with groundsel, even if they refuse all other food (they're clearly not related to the rabbits that invade *my* garden, which eat everything *except* groundsel). But most grateful of all for the plant's leaves and bountiful seeds are caged birds, as well as numerous wild species that are free to choose their own food.

As is true of many plants that grow with enthusiastic abandon, groundsel has enjoyed a long tradition as an herbal remedy, albeit a modest one. Nevertheless, the highly respected Pliny's judgment seems to have briefly lapsed when he all but attributed magic powers to the weed. Claiming its ability to cure toothache, he wrote, "If a line is traced around [groundsel] with an iron tool before it is dug up, and if one touches a painful tooth with the plant three times, spitting after each touch, and replaces it into its original ground so as to keep it alive, it is said that the tooth will never cause pain thereafter." On the other hand, perhaps we moderns have become too cynical.

The association of iron with groundsel is a thread that runs through several old-time medicinal treatments. Smelling the plant's freshly dug roots was thought to be an instant cure of headaches, provided the roots were not dug up with a tool containing any iron. Another old herbalist claimed that the plant's healing powers worked exceptionally well on wounds caused by being struck by iron.

Whether or not these particular remedies work, the weed was formerly much used as a remedy for epilepsy, cholera, and jaundice, and considered good for stomach complaints. "The leaves of Groundsel boiled in wine or water, and drunke," explained Gerard, "heale the paine and ach of the stomacke that proceeds of Choler. . . . Boiled in ale with a little hony and vineger, it provoketh vomit, especially if you adde thereto a few roots of *asarabacca* [wild ginger]."

Culpeper recommended taking the juice in wine to provoke urine and expel "the gravel in the reins and kidneys." But what he chiefly stressed about this weed is its cooling quality. "The herb," he asserted, "preserved in a syrup, in a distiled water, or in an ointment, is a remedy in all hot diseases, and will do it: first, safely; secondly, speedily." In fact, he claimed that "the distilled water performs every thing that can be expected from its virtues, especially for inflammations or watering of the eyes." However, it is unclear why he singled out "the people in Lincolnshire" (unless they were the only ones), who used the expressed juice "externally against pains and swelling, and . . . with great success." In any case, their remedy seems to be closely related to one of Culpeper's own. To dissolve "knots and kernels in any part of the body," he suggested preparing a poultice of groundsel with a little salt.

Fifteen hundred years before both Gerard and Culpeper, Dioscorides had apparently already learned to value the use of groundsel in the form of a poultice. Gerard quotes him as saying, "That with the fine pouder of Frankincense it healeth wounds in the sinues. The like operation hath the downe of the floures mixed with vineger."

During earlier centuries in England, gout is said to have been alleviated by applications of groundsel first pounded with lard. In some English rural districts, as recently as earlier this century, a poultice of crushed groundsel leaves laid externally to the stomach was deemed every bit as effective an emetic as a cup of a strong groundsel infusion. A similar application was also considered beneficial "for the gripes and colic of infants."

Culpeper described groundsel juice as "a good purgative," but stressed that "the dose should not exceed two ounces." But in more modern usage, groundsel *tea* prepared in a weak infusion appears to be preferred for the same purpose, as well as for eliminating biliousness, because such a tea is reputed to cause neither pain nor discomfort. A strong infusion, by contrast, can act as an emetic. An infusion of groundsel is often prescribed as a gargle for a sore throat. Sweetened with honey, a weak infusion has been used for thrush in children. "The leaves stamped and strained into milk and drunk" is the recipe Gerard offered to treat "the red gums and frets of children."

In Germany, the expressed juice used to be a popular remedy for worms in children. In France especially, many women have long regarded groundsel tea as one of the most reliable remedies for female disorders. They believe it not only regulates and induces menstruation but that it also soothes menstrual pains. The suggested tea for these conditions (1 ounce fresh, or ½ ounce dried herb, brought to boil, then simmered 2 minutes in 1 quart water, and strained) is taken by them in daily doses of 3 to 4 small cupfuls.

However, the juice of groundsel is most often applied as a styptic to stanch the bleeding of minor wounds, and as a cooling remedy for insect stings. The bruised leaves are still applied liberally, as they were in bygone days, to draw boils and abscesses to a head. And, old-fashioned as is the remedy, even modern-day chapped hands submit gladly to the soothing effect of soaking in a warm groundsel infusion.

For all its uses, the whole herb can be employed, excluding the flowers that have gone to seed; the juice expressed in spring is considered the most effective. To store the herb, leaves are best picked in spring, and either air dried or placed on brown paper in an oven set at a low temperature. Both the leaves, fresh and dried, and the expressed juice have a slightly saline, bitter taste, although their effect is described as soothing. Nevertheless, the internal use of groundsel should always be sparing: Research indicates that taking the herb in large doses, or over a prolonged period of time, can damage the liver and may cause cancer.

Culpeper's writings of more than three centuries ago suggest that he already had an inkling of such a possible risk, specifically regarding the expressed juice, when he advised that "a dram of the juice is sufficient to be taken inwardly, and caution should be used so that it may not work mischief." If his advice is to be taken literally, then it is necessary to be aware that a dram represents two different measures: In apothecaries' weight, it is ⅛ ounce; in avoirdupois weight, ¹⁄₁₆ ounce. In short, moderation.

Stellaria media

Stellaria media
(CHICKWEED)

VITAL STATISTICS

Annual

COMMON NAMES	(*Alsine media*) Chickweed, Starweed, Starwort, Tongue Grass, Winter Weed, Passerina, Chick Wittles, Clucken Wort, Skirt Buttons, Stitchwort
USES	Culinary, medicinal
PARTS USED	Whole herb
HEIGHT	4 to 15 inches; trailing, tangled stems, to 2 feet long
FLOWER	White, starlike, in axils of upper leaves; March to late fall
LEAVES	Oval, stalked, succulent, pale green, smooth
ROOT TYPE	Hairlike, brittle
HABITAT	Sunny lawns, roadsides, meadows, gardens, waste places
PROPAGATION	Seeds
CONTROL	Regular cultivation, weeding

It seems inconceivable that this wraithlike, delicate European native plant could be such a garden pest in much of the world, including the northern reaches of the Arctic circle. Inevitably, having always followed human settlements, chickweed is today also considered native throughout the temperate regions of North America. In fact, its zest for life is such that it keeps seeding itself and flowering again so quickly that there is hardly a time of year when this weed does not bloom somewhere in the United States, even if it has to do so occasionally under a cover of snow.

There is no denying the charm of a border of *Stellaria media* in full bloom in early spring, of its starlike flowers nestled in the upper leaf axils at the end of the plant's loosely woven nest of much-branched, trailing stems. Their petals are pure white, cleft, and narrow, and were long ago included in sundials, or floral clocks, made of herbs. Of such a sundial, the seventeenth-century poet Andrew Marvell wrote,

How well the skilful gardener drew
Of flow'rs and herbs this dial new;
Where, from above the milder sun,
Does through a fragrant zodiack run,
And, as it works, th' industrious bee
Computes its time as well as we!

By so placing various plants that a different flower opened or closed at each hour of the day, many a patient gardener created his own colorful, fragrant timepiece with which to reckon "sweet and wholesome hours." Today, as they always have—on clear days only, however—chickweed's stars turn their bright little faces toward the sun from about nine o'clock in the morning until the petals close again. On rainy or overcast days, the flowers remain furled.

The plant's leaves and stems are pale green and succulent, the smooth, ovate leaves growing in pairs, on short, flat stalks below, stalkless above. What adds an uncommon feature to the stems is the fine line of hairs that runs along them on only one side at a time, from one leaf pair to the next; at the second leaf joint, the hairline continues on the opposite side of the stem to the third joint, alternating in this manner along the full length. And every night, when the flowers begin to close, the plants enter what is termed the "sleep of plants": like kitten paws, each pair of leaves folds over the tender new shoots and buds above it until the warmth of the following day wakes them again. Several plants go by the name chickweed, including the dark green woolly and whiskery lawn invader *Cerastium vulgatum*, commonly known as mouse-ear chickweed. However, *S. media* is generally re- garded as the "real" chickweed. In the days of Nicholas Culpeper, it was known as *Alsine media*, a name by which the weed often continues to be listed, either alone or together with *S. media*. To confuse the name issue still further, and to test the patience of the unwary, whether the chickweed species is named *Stellaria*, *Alsine*, or *Cerastium*, all of them are members of the large Pink family, the Caryophyllaceae.

Its impressive family connections notwithstanding, chickweed is often treated as a major nuisance in the home garden, be it in lawns or in flower beds. It does tend to grow everywhere, although it prefers to sprawl over cultivated soil left undisturbed by such human intervention as weeding. In my own garden, it usually hugs the periphery of the vegetable bed, and reaches up from amid such border perennials as lavender, candytuft, or *Phlox subulata*.

Yet, profuse and prolific as it is, chickweed has the decency at least to be shallow rooted and is, therefore, easily removed. Through diligent weeding, as well as regular cultivation, the plant can be quickly eradicated from flower and vegetable beds. An effective method for freeing lawns of chickweed is through frequent mowing, thereby successively preventing, or at least curbing, the weed's formidable reproductive capabilities. Given free range, chickweed can produce five generations of chickweedlets in a single season! A more drastic approach is an early-morning dusting of ammonium sulphate on the weed, while dew is still on it. Not only does this burn and kill the intruder, it also acts as a growth stimulant to the grass roots. A word of caution, however: This treatment not only kills the weed but may do likewise to the grass tops.

There is also a fourth method for chickweed control, and that is to eat it. At least, that's my solution. Especially in early spring, often long before spinach or lettuce can be even planted, let alone harvested, there is already an abundance of chickweed. Its tangled mass—not unlike a mop of unkempt curly hair—may daunt a first-time reaper seeking to extract individual stems and leaves. However, because the entire herb is edible, harvesting it consists of nothing more than gently seizing a handful and pulling it free. If you intend to cook chickweed, it is wise to pick at least double the amount you may deem necessary—like spinach, it cooks down to a fraction of its fresh volume.

Chickweed eaten fresh adds a delightful piquancy to salads, as a change from watercress, for instance. Or it can be used as the main salad green. Chickweed also makes a tasty and nutritious addition to my sandwiches, in place of lettuce. Chopped, it can be added to chicken or egg salads; it gives these traditional standbys a zesty newness. Or a pinch of chickweed atop cold soup or salmon mousse makes an attractive summer garnish. Steamed or boiled for 3 to 5 minutes, and lightly seasoned with pepper, salt, and lemon juice, chickweed ranks among my favorite spinach taste-alikes. First steamed, it can also be used as a filling for omelets or as a low-calorie stuffing for fish or fowl. According to Gerard, even "little birds in cadges (especially Linnets) are refreshed with the lesser Chickweed when they loath their meat, whereupon it was called of some 'Passerina.' "

Although modern herbalists since Gerard's time, more than four hundred years ago, have held chickweed in high regard for its medicinal properties, their ancient counterparts appear to have completely ignored the weed. It may never have ranked among the *major* healing herbs known to man, yet it used to be greatly valued in the treatment of blood toxicity and of so-called hot diseases, such as fevers and inflammations, be these internal or external. Described as both emollient and refrigerant, chickweed is rich in potassium salts and can be used both fresh and dried.

Mostly, chickweed is used in the form of an ointment. This practice was already favored in the time of John Gerard, who explained that "the leaves of Chickweed boyled in water very soft, added thereto some hog's grease, the powder of Fenugreeke and Linseed, and a few roots of Marsh Mallows, and stamped to the forme of Cataplasme or pultesse, taketh away the swelling of the legs or any other part . . . in a word it comforteth, digesteth, defendeth and suppurateth very notably."

Culpeper, offering help for "sinews when they are shrunk by cramps or otherwise," recommended the following recipe: "Boil a handful of Chickweed and a handful of dried red-rose leaves, but not distilled, in a quart of muscadine, until a fourth part be consumed; then put to them a pint of oil of trotters [pig's feet], or sheep's feet; let them boil a good while, still stirring them well, which being strained, anoint the grieved part therewith warm against the fire, rubbing it well with your hand, and bind also some of the herb, if you choose, to the place, and with God's blessing it will help in three times dressing."

In home medicine, chopped chickweed is often simmered in lard or another setting fat to prepare a soothing ointment for hemorrhoids, as well as for skin diseases and sores. The same ointment has also been applied to inflamed eyes.

Boils, carbuncles, and abscesses are said to respond quickly to the direct application of poultices made from fresh chickweed. Or, the fresh or dried herb is infused, and the tea taken internally, while the herb is placed on the affected part. Similarly, drinking the freshly extracted juice of chickweed has been recommended, or swabbing the infected part with it—or doing both—in cases of minor skin infections or eruptions.

For a number of other afflictions, Culpeper advised the following: "The herb bruised, or the juice applied, with cloths or sponges dipped therein to the region of the liver, and as they dry to have fresh applied, doth wonderfully temper the heat of the liver, and is effectual for all impostumes and swellings whatsoever; for all redness in the face, wheals, pushes, itch or scabs, the juice being either simply used, or boiled in hog's grease; the juice or distilled water is good use for all heat and redness in the eyes . . . as also in the ears."

Although chickweed tea is available from herbalists, a decoction can be prepared by simmering an ounce of the fresh herb in a pint of water for 20 minutes. Such a tea has long been recommended for coughs, colds, hoarseness, and hemorrhoids, as well as sore eyes and arthritic conditions. Herbalists consider it useful in cleansing the blood and ridding both liver and kidneys of harmful wastes. They also prescribe it as a tonic to the weary and to restore strength to the weak. Not least, because of the mildly diuretic and laxative properties ascribed to it, chickweed has also been used as an aid in weight reduction.

The trouble is that by the time I had discovered all these many uses for chickweed originally, I had banished most of it from the garden, scorning it as a plant beneath my contempt. And now I found myself in the embarrassing position of having to encourage it to resume its tenancy—under improved terms and conditions, of course. Luckily—that is, typically—it did so with most forgiving alacrity.

Taraxacum officinale

Taraxacum officinale

(DANDELION)

VITAL STATISTICS

Perennial

COMMON NAMES	Dandelion, *Leontodon taraxacum*, *Taraxacum Densleonis*, Priest's Crown, Swine's Snout, Milk Gowan, Doonheadclock, Pissabed, *Pissenlit*, Monk's Head, *Herba Urinaria*, *Herba Taraxacon*, Irish Daisy, Puffball, Peasant's Cloak, Yellow Gowan, Blowball, Telltime, Lion's Tooth, Heart-Fever-Grass
USES	Culinary, medicinal, cosmetic, commercial
PARTS USED	Roots, leaves, flowers
HEIGHT	2 to 12 inches
FLOWER	Golden yellow, 1-to-2-inch diameter, solitary atop smooth, hollow stem; open in full daylight, closes at dusk; pappus white; March to September
LEAVES	1 to 18 inches long, jagged, grooved, bright green, hairless; emerge directly from root, forming rosette on soil surface
ROOT TYPE	Taproot, 2 to 12 inches long, fleshy, sometimes forked, wrinkled brown externally, white internally, with pale yellow center
HABITAT	Lawns, fields, meadows, waste places, roadsides
PROPAGATION	Seeds, roots
CONTROL	Digging up roots, mowing before flowers open

About 4 lb. of seed to the acre should be allowed, sown in drills, 1 foot apart. . . . The yield should be 4 or 5 tons of fresh roots to the acre in the second year."

If the start of this chapter had not already given the game away, I wager you would be eagerly waiting to read the name of such a remarkable crop. Instead, I hear you gasp incredulously, "an acre of *what?*" Yes, I refer to the very same dandelion whose bright golden flowers ruin your otherwise perfect lawn in spring; the same weed you spend hours every year painstakingly pulling and digging up; the weed for which garden supply houses stockpile whole arsenals of deadly weapons. In the United States especially, probably no other plant is so well known, so easily recognized, so much hated, so systematically singled out for extermination—and so little understood—as the dandelion.

In fact, however, dandelions are good business—even in the United States. Every year, large quantities of the plant's leaves supply a considerably popular demand for fresh spring greens in many ethnic grocery stores and supermarkets. Added to the domestically grown dandelion roots, the United States also annually imports more than 100,000 pounds for use in patent medicines. Not bad for a weed.

For at least 1,000 years, the dandelion has been in constant use as both a food and a medicine. Like so many plants, its origins were in the Mediterranean regions of Europe and Asia Minor. The ancient Egyptians are said to have known and used the plant; Theophrastus described it some 300 years before Christ. But it was the Arabian physicians of the early Middle Ages who first officially recognized the plant's medicinal properties and named it *Taraxacon*, from the Greek *taraxos* for "disorder," and *akos* for "remedy."

Some 300 years later, in the thirteenth century, allusion to the plant appears in Welsh medicines. A German monograph, written in 1485, refers to the name by which the weed is known in most European languages. It seems that a surgeon, Wilhelmus, who greatly admired its virtues, compared it with "eynem lewen zan, genannt zu latin Dens leonis" (a lion's tooth [*Löwenzahn* in modern German], called in Latin *Dens leonis*). The insistent reference to a lion's tooth has been variously attributed to the shape of the weed's leaves, or its roots, or its flowers. In the course of time, *Dens leonis*, the former specific Latin name, led easily enough to the Greek *Leontodon*, the genus to which Linnaeus assigned the plant; from there it was a short hop to the French *dent-de-lion*, which, in crossing the rough waters of the English Channel, became corrupted only slightly to dandelion. And sometime after Linnaeus, the German professor Georg Heinrich Weber gave the genus its permanent botanic name.

In the days of Gerard, Parkinson, and Culpeper, English apothecaries knew the plant as *Herba Taraxacon* and as *Herba Urinaria*. In time, it earned other names as well. The Irish used to call it heart-fever-grass, probably because of its ability as a bitter to ease heartburn. When the mature flowerhead closes up, it resembles the snout of a pig, or swine's snout. Blowball and telltime remind the child in us what fun it is to blow on, or to watch a breeze carry off, the dandelion seeds on their plumed parachutes. And when all the seeds are gone, all that remains is a kind of naked pate, just like the shorn heads of monks and priests in medieval times. But the name that deals most unceremoniously with our *Herba Urinaria*, for its diuretic effect, is the French *pissenlit*, or pissabed in English.

Like the Arabian physicians before him, Gerard likened dandelions to wild chicory and observed (as we know all too well!) that dandelions "floure most times in the yeare, especially if the winter be not extreme cold." Culpeper found it "very effectual for the obstructions of the liver, gall, and spleen," for jaundice, urinary infections, and consumption. But he concludes with a sudden outburst of anger against English physicians. "You see here," he writes, "what virtues this common herb hath, and that is the reason the French and Dutch so often eat them in the spring; and now if you look a little further, you may see plainly without a pair of spectacles, that foreign physicians are not so selfish as ours are, but more communicative of the virtues of plants to people."

Over the centuries, no temperate zone north of the equator escaped the dandelion; it can turn up in the most unlikely places. Only last summer, I discovered a Lilliputian version in radiant bloom, snugly tucked under an overhanging rock on a Swiss alp, at an altitude of more than 10,000 feet. The flower was no larger than a thumbnail, the stem a scant half-inch, hunched into the protective rosette of leaves that resembled a kitten's rather than a lion's teeth. Perhaps a mountaineer had planted it there deliberately, just as the early European immigrants deliberately introduced it to the New World, to heal and comfort them in times of need. Native Americans, too, quickly learned the value of a tonic tea prepared from the leaves or roots. As the pioneers pushed westward, they encouraged the good-natured workhorse of a weed to excel in yet another occupation, as a food source for bees. Even today, dandelion ranks high among honey-producing plants, thanks to its bounteous stores of pollen and nectar. In fact, with what is surely an angel's patience, it has been observed that no fewer than ninety-three different kinds of insects help themselves to the dandelion's lavish larder.

It is difficult to say in which capacity dandelion is better known or valued, as food or medicine. In fact, because it is virtually impossible to separate the two when it comes to this weed, I will do what much of the existing literature about the dandelion does, which is to meander between them. A word of caution, however: Should you decide to try any of the dandelion foods, drinks, or remedies discussed here, be sure you do not use the dandelions from any lawns or fields that have been subjected to herbicides or other chemical contaminants.

Certainly in southern Europe the plant's values as a food or medicine have always been regarded as one and the same. There, it is not at all uncommon, even now, to see families sally forth into the fields in spring, armed with baskets and sharp knives, to harvest the tender young dandelion greens. So rich are these in vitamins that, whether they are eaten as zesty, refreshing salads, cooked as a potherb, or infused as a tea, they act as a tonic on the winter-weary body, and purify the blood. In upstate New York, too, in spring, many men and women with baskets or bags, and a knife in hand, can be seen along roadsides, bottoms up in the air, gathering dandelions. Even in New York City years ago, I remember Sundays in early spring, when my Italian neighbors, all of them together in a group, used to take the train to Long Island or New Jersey. Amid much laughter, chatter, and food, they herded their toddlers and teenagers on a special day's outing to what were then open fields. And when they returned home, in time to cook supper, their baskets were full of dandelion leaves and roots.

The best dandelions are found where no lawnmower has touched them, where the soil is rich and loose. In such a place, the weed can be dug up fairly easily with a fork or spade, and the taproot is likely to be a juicy single, the thickness of a finger. The poorer the soil, the more forked is the root, and the less desirable, because it tends to be tough. (Remember, as with most taprooted plants, if a piece is left in the ground, a new plant will grow from it.) The top of the root is generally 1 to 3 inches below the surface of the soil, and forms a crown of blanched leaf stalks. Nature provides few finer delicacies.

The crown is severed from the root, deeply enough so that the blanched leaf stalks remain together. Leaves that have turned green are discarded, and the crowns thoroughly washed. To prepare a small salad for two persons (most of my dandelion recipes come from friends), I thinly slice 8 to 12 crowns crosswise, add a small finely chopped onion, a pinch of sugar, salt and pepper to taste, a little olive oil and cider vinegar. For the more robust flavor of a recipe called *pissenlit au lard*, I first blanch, then grill a couple of bacon rashers, then crumble them on a raw dandelion salad dressed with vinegar and olive oil, seasoned with pepper and salt, and garnished with finely chopped parsley, borage, and chives.

As a vegetable, the crowns or the very young rosettes of leaves can be boiled or steamed for 3 to 5 minutes, drained, and seasoned with lightly browned butter, lemon juice, pepper and salt, and chopped parsley. For a simple, nutritious, not to mention inexpensive lunch, I prepare dandelion greens like creamed spinach, served on whole wheat toast. Young dandelion leaves are also delicious on ham and cheese sandwiches. Unfortunately, as soon as a dandelion blooms, its leaves become too bitter for most palates.

Ideally, the roots to be used either as food or medicine should not be more than one year old. Some people say they are best dug in September and October. To prepare them for dinner, they are usually washed, peeled, and sliced crosswise. Most people are particularly fond of them steamed like parsnips. Or, after being steamed, the roots can also be quickly tossed in melted butter browned to a nutty flavor and seasoned with pepper, salt, chopped parsley, and a squeeze of fresh lemon. During his travels in the New World, Swedish botanist Peter Kalm observed as early as 1748 that French Canadians prepared spring-dug roots of dandelion as a bitter salad. Icelanders fry their roots.

If the roots are to be dried—a process that can take several hours—they are generally washed thoroughly, sliced unpeeled, and slowly roasted in an oven set at about 180 degrees Fahrenheit until they are completely dried and tinted a medium brown, then stored in a tightly closed container. When needed, they can be decocted into an herbal tea; ground into powder, they can be prepared as a coffee substitute that is wholly pleasing and beneficial to the entire system. And to create a kind of mocha drink in reverse that does not keep a person awake, many dandelion fans add a teaspoonful of ground dandelion root to their hot chocolate on a cold winter's night.

It's the dandelion flowers that pack a wallop—literally. If you have never tasted dandelion herbal wine, it is one of the most elusive, delicately fragrant flavors imaginable, the color pure liquid gold. My sister used never to let a spring go by without making at least one gallon of this intoxicating, *healthy* tipple for holiday and other family gatherings.

Together with our children, when they were still young, Helen and I first gathered 6 quarts of flowers on a warm, dry spring day, and placed them in a clean 6-gallon crock. It is important to note that the flowers should not be washed. Next, Helen filled the crock with fully boiling water and allowed the blooms to steep overnight, covered with muslin. The following day, she strained the liquid through the muslin and returned the liquid to the crock (discarding the flowers). To the liquid she added 4 sliced lemons, 4 sliced oranges, two 12-ounce boxes black raisins, 2 cakes yeast, and 6 pounds sugar. She stirred together all ingredients, covered the crock with a lint-free towel, and set the crock in a warm, draft-free location, stirring the contents once daily for seven days, or until the bubbles stopped rising, always careful to skim off the scum. Then she left the contents undisturbed for one day to allow the sediments to settle. On the ninth day, she siphoned the wine into cork-stoppered bottles and allowed it to mature at least six months. By Christmas, the dandelion wine, which is traditionally considered one of the finest tonics, was ready for us—and we for it. Of course, the brew was far better still after ripening for three or five years!

Not many weeds can claim that every part of them is known to benefit humans. The dandelion, however, is one of these, thanks to its high content of protein, calcium, phosphorus, iron, riboflavin, niacin, and vitamins A, B_1, and C. Because no part of the plant is poisonous, herbalists recommend it freely. Its most common preparations are either as an infusion or a decoction, usually made of the root, although the leaves are also used. The reason the roots are preferred is that they are considered to be the major storehouses of the plant's attributed beneficial constituents.

To prepare an infusion, a cupful of boiling water is poured over 1 ounce of the finely chopped fresh or dried root, and steeped, covered, for 30 minutes. For the decoction, the mixture is simmered for 20 minutes. In both cases, honey is usually added. Dandelion tea, according to Mrs. Grieve, is "efficacious in bilious affections"; in fact, however, most herbalists believe that the beneficence of such a tea is almost boundless.

For a century, dandelion was regarded as an official drug in the United States, and the dried root remains listed in the *U.S. Pharmacopoeia*, although only its use as a tonic is substantiated. Nevertheless, it is an established fact that dandelion extract is capable of greatly increasing the volume of bile, a necessary factor in the treatment of some liver and kidney ailments.

Among herbalists the world over and throughout history, the dandelion has been regarded as an excellent treatment for ailments of the liver, kidneys, gall-bladder, pancreas, and blood, usually prescribed in the form of a decoction made from the roots (2 ounces root or leaves in 1 quart water boiled down to 1 pint). It has been prescribed for edema, jaundice, gout, eczema, cellulite, and "heat in the joints." It has been credited with stimulating the circulation, cleansing the blood, aiding sluggish digestion, relieving constipation and diarrhea, curing stomachaches, calming the nerves, and eliminating cholesterol. Even such divers conditions as arteriosclerosis, obesity, catarrhal conditions, and arthritis are said to have responded well to dandelion. Accompanied by a healthy, balanced diet, dandelion root tea is considered particularly beneficial in cases of hypoglycemia and diabetes.

An eighteenth-century herbalist described dandelion as "a blessed medicine," and in these terms it has been valued through the centuries, wherever it grows. In China, it is known as a blood cleanser, a tonic, and a digestive aid. In India, it has been traditionally relied upon to heal liver complaints. The root, dried and coarsely ground, was said to heal snakebites. In European countries, the sticky, milky juice of the entire plant has long been used to stimulate glandular activity. In former times, the juice was also prescribed for malignant growths. Also, many people make frequent applications of it to warts, until the latter turn black and drop off.

In Germany, kidneys no less august than those of Frederick the Great were treated with an extract of decocted dandelion root. Germans have also tradition-ally so much respected dandelion in the treatment of eye ailments that they sometimes refer to the weed as "eye root." Specifically, they consider the milky juice of the stem and root of greatest benefit in effecting clear eyes. Certainly, the juice extracted from the spring-dug plants remains a main feature of the so-called blood cleansing spring cures at some German spas.

In England, John Evelyn noted as early as the sixteenth century that dande-lion leaves "have been sold in most *Herb Shops* about *London* for being a wonderful Purifier of the Blood." Just as they did then, for treating chronic liver congestion, some English herbalists still prescribe the daily consumption of a broth made by stewing sliced dandelion roots in water, adding a few sorrel leaves and an egg yolk. To stimulate the appetite and promote digestion, or to soothe an irritated stomach, a decoction of dandelion root is regarded as valuable. Similarly, the decoction has always been taken as a remedy in cases of eczema, scurvy, and other skin eruptions.

Although it does so on a small scale, dandelion even plays a role as a beauty aid. With a handful of dandelion flowers infused in a pint of water and strained, European women already centuries ago used to rinse their faces morning and night to get rid of freckles. I personally prefer to use it as a facial steam. In England today, a strong dandelion infusion often serves as a tonic facial rinse, or as a refreshing addition to an herbal bath. Dandelion can even be employed as a dye. The flowers are used to make a yellow dye for wool; the entire plant produces a rich magenta.

In the course of time, as dandelion leaves gained recognition as a special ingredient in digestive and diet drinks, they also found their way into so-called herb beers. Whether dandelion beer evolved specifically for reasons of health or economy is a moot point. At any rate, it won great favor in numerous rural and industrial centers, both in England and in Canada, and is described as a whole-some fermented drink. A neighbor of ours, Mr. Moll, used to make it annually. He swore by it and let me taste some once when I was quite young. I didn't like it. But then, what did *I* know about such things at the time?

In any case, according to Mr. Moll, you pull up ½ pound whole dandelion plants as soon as they begin to bloom in spring, wash them thoroughly, and remove the hairs from the root. For 15 minutes you boil the plants with ½ ounce cubed, crushed gingerroot and the peel (minus pith) of 1 lemon in 1 gallon water. Then you strain the boiled liquid over 1 pound light brown sugar and 1 ounce cream of tartar in a 2-gallon kettle or crock, stirring until the sugar is dissolved. Cool the mixture to lukewarm, then add the juice of the lemon and 1 ounce yeast, cover the crock, and let it stand in a warm room. After 3 days, strain the liquid, decant it into screw-topped bottles, which are stored on their sides for about a week. As soon as there is a hiss when a bottle top is loosened, the beer is ready. Incidentally, Mr. Moll claimed that dandelion beer doesn't keep very long—isn't that a bit of luck?

Not too surprisingly, confronted with a plant that is endowed with such a diversity of powers, somebody was bound to attribute magical properties to it. Sure enough, the sixteenth-century Matthiolus recorded that "magicians say that if a person rub himself all over with [dandelion], he will everywhere be welcome and obtain what he wishes."

Alas, the only power that seems to have eluded the dandelion through all the centuries and across thousands of miles is the power to win the otherwise so generous American hearts.

Tussilago farfara

Tussilago farfara

(C O L T S F O O T)

VITAL STATISTICS

Perennial

COMMON NAMES	Coltsfoot, *Filius ante Patrem* (Son-Before-the-Father), Coughwort, Baccy Plant, Donnhove, Bullsfoot, Foalswort, *Pas d'Ane* (Ass's Foot), Clayweed, Horsehoof, Fieldhove, Clutterclogs, Sweep's Brushes, Wild Rhubarb
USES	Culinary, medicinal, cosmetic, household
PARTS USED	Whole plant
HEIGHT	Up to 8 inches in flower
FLOWER	Golden yellow, similar to dandelion
LEAVES	Hoof shaped, slightly toothed, 4 to 8 inches diameter
ROOT TYPE	Fleshy, underground rootstocks
HABITAT	Roadsides, waste places; sandy, clayey soils; prefers full sun
PROPAGATION	Root division; fluffy seedheads
CONTROL	Regular weeding
CONTRAINDICATIONS	Caution: Internal use may be unsafe.

Of *Tussilago's* numerous common names, *coltsfoot* is probably the most representative and descriptive. However, it does not prepare the uninitiated for its inherent botanic surprise, namely, its two-stage development. Coltsfoot blossoms appear singly, in early spring, layered rays of golden yellow atop succulent stems that are covered with pinkish woolly scales. Only after the flowers have faded and turned into seedheads crowned with tufts of silky down, called *pappus*, do the plant's distinctive leaves emerge from the base. Long stemmed, large, and hoof shaped, these toothed leaves may reach a width of 8 inches, their upper surfaces a glossy green, while the undersides are covered with loose, white, felted hairs.

It is this two-stage development, the appearance of flowers before leaves, that earned coltsfoot one of its earliest known sobriquets, *filius ante patrem*. Not surprisingly, on first encountering the plant, many of the early botanists, including even Pliny, were duped into believing that the coltsfoot had no leaves at all. For this reason, many books separately illustrate the plant's flowers and leaves.

For more than 2,000 years, coltsfoot has been prescribed, prepared, and used as one of history's most popular cough remedies. Ancient Greeks called the plant *bechion*, or "cough plant." The earliest known mention of it as a healing plant appears to come from Hippocrates, the "father of medicine," born in 460 B.C., who recommended it, combined with honey, as beneficial in treating ulcerations of the lung. Certainly, the weed's botanic name is based on the Latin *tussis*, "cough," and *agere*, "to drive."

There is less assurance about the origins of the weed's specific name, *farfara*, and most references avoid defining it. Mrs. Grieve, however, attributes it unequivocally to "*Farfarus*, an ancient name of the White Poplar, the leaves of which present some resemblance" to this plant. Another definition relates *farfara* to *farina* because of the whitish fuzz covering the undersides of the leaves. This definition might well be supported by *farfarris*, the Latin word for "spelt," a kind of wheat. Of course, it is also possible that the answer lies in the Spanish *en farfara*, meaning "immature, half done, unfinished," in possible reference to coltsfoot's two-stage development.

Originally a native of Asia, North Africa, and Europe, coltsfoot was introduced to North America by early European immigrants. From the coastal regions, coltsfoot has steadily advanced inland over the centuries, particularly in the Northeast. Here, even before the last winter snows and ice have melted, coltsfoot is apt to unfurl its blossoms under the warm spring sun, looking like nothing so much as scattered gold coins glittering among last year's matted leaves. Because it can thrive even in the poorest soils, coltsfoot is often found in waste places, as well as on sunny roadsides and embankments, near shallow brooks, or along the edge of woods. In fact, it may be said that coltsfoot is a confirmed sun worshiper. Its flowers open only when the sun shines fully on them; they stubbornly refuse to unfurl on overcast days, just as they routinely close up at dusk.

The weed especially likes sandy and clayey soils, sending out a formidable root network from which arise numerous flower stalks. So dense is this root network that few other plants can grow in its immediate vicinity. For this reason, coltsfoot is an ideal cover plant for otherwise inhospitable garden locations, where soil is sparse and poor. Of course, for the very same reason, it would be headlong recklessness to introduce the coltsfoot into formal flower beds. The coltsfoot's audacious underground root network is extremely difficult to extirpate, because not only do the succulent roots break easily but any pieces left in the soil will generate new plants. Who knows, perhaps this is how this plant earned the nickname *clutterclogs.*

For one purpose or another, all parts of the coltsfoot can be used—even the *pappus.* Goldfinches and chickadees favor it for lining their nests, just as Scottish Highlanders in the past often gathered this fine down to stuff their pillows and mattresses. And in the days before matches were invented, when tinderboxes were still popular, the felted hairs that are easily rubbed off the undersides of the leaves were collected and "wrapped in a Rag, boyled a little in Lee [lye?], adding a little [solution of] Salt-Petre, and dried in the Sun." At least, so says John Pechey in his *Compleat Herbal.*

In its capacity as a reputed demulcent, expectorant, and tonic, coltsfoot has been regarded as "nature's best herb for the lungs and her most eminent thoracic" by the great medical authorities of successive ages. Said to abound in mucilage, coltsfoot has been used most often as a tea, a syrup or candy, as poultice, gargle, enema, and tobacco, as well as a cosmetic aid. Coltsfoot's importance as a medicinal plant was such over the centuries that painted signs of it became the symbol of apothecaries in Paris, who displayed it above, beside, or on the doorways to their shops.

Controversial by modern standards is the long-held advocacy of smoking dried coltsfoot leaves to relieve coughs, asthma, and bronchial congestion. Nevertheless, this practice seems to have won the approval of both ancient and more recent herbalists, from Dioscorides, Pliny, and Galen to Gerard, Culpeper, and Linnaeus, and continuing at least into the recent past, when coltsfoot formed the basis of the British Herb Tobacco. The following recipe for an herbal tobacco appeared in the October 1953 issue of the British publication *Health from Herbs:* "Mix thoroughly 2 parts coltsfoot, ½ part camomile flowers, ½ part thyme, 1 part eyebright, 1 part betony, and ½ part rosemary."

Pliny suggested mixing dried roots with the coltsfoot leaves. Gerard advised that "the fume of the dried leaves taken through a funnell or tunnell, burned upon coles, effectually helpeth those that are troubled with the shortness of breath, and fetch their winde thicke and often." Culpeper believed that smoking coltsfoot tobacco was good for people "who have thin rheums and distillations upon their lungs, causing a cough," but suggested that those suffering from "a hot, dry cough, or wheezing, and shortness of breath" should use "the fresh leaves, or juice, or syrup thereof."

The eighteenth-century Linnaeus observed that his Swedish compatriots smoked coltsfoot tobacco for its curative values, whereas during World War I, many an English soldier substituted coltsfoot for a real smoke. In fact, some modern herbalists believe that herb tobaccos may help people give up addictive tobacco. Reputedly, one of the preferred mixtures consists of the following dried herbs: 1 pound coltsfoot, ½ pound each eyebright and buckbean, 4 ounces wood betony, 2 ounces rosemary, 1½ ounces thyme, 1 ounce lavender. To these, some people like to add camomile flowers and rose petals. Those who like a mild tobacco simply increase the proportion of coltsfoot.

Whether or not smoking coltsfoot is deemed advisable, I clearly remember a

Bavarian neighbor long ago, a weatherbeaten, crusty old codger who did not consider a day complete without his stein of *Dunkles* (stout) and his pipe smoke of *Huflattich*, which translates coltsfoot as "hoof lettuce."

There is no doubt that coltsfoot has ranked high, throughout its long history, as a preferred treatment for coughs and colds and all lung disease. Because of its high mucilage content, the plant was also considered a soothing demulcent on inflamed and irritated mucous membranes. Although all parts of the weed are used, it is predominantly the flowers and leaves that are gathered early in their respective seasons, the leaves being generally preferred, fresh or dried. Infusions of the plant should always be strained through cheesecloth to remove the fine down on the leaves, which otherwise can cause irritation to the mucous membranes.

The American Indians as well as early settlers ate coltsfoot roots, raw or boiled, to treat coughs. With the leaves they made soothing teas, and to reduce severe congestion, they wrapped the patient in blankets soaked in a hot coltsfoot infusion.

German homeopaths today continue to believe in the efficacy of enemas of coltsfoot infusions for enteritis. Similarly, herbalists throughout the centuries seem to have agreed that drinking one cupful or more daily of an infusion or decoction of coltsfoot alone, or combined with other herbs, and sweetened with honey, is beneficial not only in ailments of the lung and throat, but also in cases of gastritis and enteritis. (Other herbs might include horehound, hyssop, ground ivy, and marshmallow.)

Many British World War I veterans suffering from lung disease due to poison gas were treated with numerous cups of honey-sweetened tea made of coltsfoot and comfrey leaves. Coltsfoot tea is also said to be a soothing tonic. And a coltsfoot infusion made from fresh or dried leaves and used as a gargle is credited with having a calming effect on a sore throat. In spite of so many recommendations, however, it should be noted that some recent laboratory tests on rodents have indicated that the prolonged or excessive internal use of coltsfoot may cause cancer.

Among long-standing domestic cough remedies in England are syrup of coltsfoot and coltsfoot rock, a hard candy or cough drop. To prepare the syrup, which used to be administered for chronic bronchitis, the *British Pharmacopoeia* suggests using the flower stalks: 1 pint strong infusion is strained into an enameled saucepan, and 2 pounds sugar dissolved in it over low heat; the mixture is gently boiled for 10 to 15 minutes, then strained through cheesecloth and stored in screw-top bottles.

For the cough drops, the syrup is gently boiled until it forms a hard ball when it is dropped from a teaspoon into ice-cold water. The lozenges are said to be soothing for both coughs and sore throats. Another recipe suggests using the slender, creeping rootstock cut into 1-inch pieces. I know my English grandmother used to keep them in a small tin, although she bought them from the druggist.

Warm compresses wetted in a strong infusion of fresh or dried coltsfoot and elder flowers are recommended for easing inflammations, swelling, and burning of the skin. Culpeper wrote that the distilled water of coltsfoot "is singular good to take away wheals and small pushes that arise through heat; as also the burning heat of the piles, or privy parts, cloths wet therein [the distilled water] being thereunto applied." Modern-day herbalists prepare poultices from crushed coltsfoot leaves and roots, mixed with a little honey to form a paste. The poultice is applied to badly healing wounds, skin ulcerations, and erysipelas (St. Anthony's fire). In France, in times gone by, the poultices were also applied to ease scrofula, a tuberculous disease of large glandular swellings also called the "king's evil." This, it was commonly believed, could be cured by a touch of the king's hand on the day of his coronation.

Cosmetically, coltsfoot has always played a minor, but nevertheless useful role. A simple infusion of the leaves makes a refreshing face lotion. Specifically, it is said to reduce facial thread veins. A compress dipped into a warm infusion and applied to the face is considered soothing to boils, acne, or spots. Whole coltsfoot leaves that have been steeped briefly in boiling water and sufficiently cooled not to cause discomfort have been put to similar use as a poultice.

For a warm poultice, a handful of fresh leaves, crushed or finely chopped, is added to enough water to make a thick pulp, then heated briefly in a double boiler, and allowed to cool sufficiently not to burn the skin, before being applied. By contrast, a poultice that has been chilled for a half hour before being applied to the eyes has been suggested by some herbalists for reducing or even eliminating that morning-after or sleepless-night puffy look. A little added honey is said to make the poultice even more soothing.

Although coltsfoot has never won even a passing glance from epicures, it was in the past occasionally included in modest dinner fare. On such occasions, the young leaves were a part of a green-leaf salad, or added to soups, or they were boiled or steamed as a green vegetable. In their most elaborate preparation, coltsfoot leaves were dipped in batter and fried, to be served with a garnish of sharp, spicy mustard. However, even this limited culinary usage appears to have died out.

The same does not seem to be true of a beer called cleats. Some rural Englishmen even today enjoy "a pint of bitter" of another kind than is normally served in pubs. The brew is probably best known up north, in Yorkshire, where the coltsfoot plant is often called cleats. William Smith's recipe suggests the following: Place into a muslin bag 1 handful each of coltsfoot, stinging nettles, dandelion, and hops, plus 1 ounce ground fresh gingerroot. With the top of the bag tied, drop it into 1 gallon boiling water; keep at a gently rolling boil for 20 minutes, stir in 2 pounds sugar until dissolved, then cool to room temperature. Add 1 ounce fresh yeast, cover the crock, and let it stand in a warm place for 3 days, until the liquid is fermented. Strain the brew into screw-topped bottles; it will be ready to drink in another 8 to 10 days.

Although I have not personally tasted this drink, I have it on the authority of an imbiber emeritus at a pub in the wilds of northern England that cleats is not only a pleasant-tasting tipple but a "dem healthy one." So, cheers! But please keep in mind that coltsfoot is first a medicinal plant. Like all medicines, it should be consumed in moderation, and in view of recent scientific findings, it may best not be consumed at all.

Urtica dioica

Urtica dioica
(STINGING NETTLE)

VITAL STATISTICS

Perennial

COMMON NAMES	Nettle, Stinging Nettle, Common Nettle
USES	Culinary, medicinal, cosmetic, commercial, household
PARTS USED	Whole herb
HEIGHT	3 to 5 feet
FLOWER	Tiny, greenish, clustered in leaf axils; July to September
LEAVES	Heart shaped, toothed, dark green, hairy
ROOT TYPE	Rhizomic
HABITAT	Roadsides, waste places, hedgerows
PROPAGATION	Seeds, divisions
CONTROL	Digging up rhizomes

It is probably safe to say that most people would be reluctant to wax eloquent about stinging nettles. After all, the name says it all, doesn't it? Alas, as happens so often in the annals of weeddom, the answer must be circumlocutory. It is simply impossible to ignore a plant that is capable of feeding, clothing, and healing the human body; a plant that has proved commercially successful, in this century alone, in at least two different industries; a plant that has been in continuous use for no fewer than 3,000 years.

However, nobody could guess all this from merely looking at the plant; it commands little attention, unless it is touched. As is true of the entire *Urtica* tribe of nettles—there are some 500 species, mainly tropical, around the world—the name of the genus is rooted in the Latin *urere*, "to burn," and that little word is surely the horticultural understatement of all time. If you have ever been stung by a nettle, you know what I mean.

Unless protectively clothed and gloved, nobody passes the nettle's formidable weaponry unscathed. All parts of the plant, the single-stalked stems and the toothed, dark green, heart-shaped leaves alike, are downy and covered with myriad stinging hairs. Each sting is a minuscule sharp and hollow needle arising from a swollen base that contains the venom. Even if you merely brush lightly past the plant, the pressure is enough for the stings to pierce the skin and for the venom to be instantly released. Within a split second, you feel the resultant burning sensation. I hasten to add that traditional antidotes for immediate relief from nettle stings are, incredibly, the expressed juice of nettle itself, or else the application of crushed leaves of curled dock, a weed that generally grows near nettles. Plantain also acts as a soothing antidote. As a preventive measure, the burning property of nettles is quickly destroyed with heat, as when the leaves are dried, or when they are boiled or steamed. In fact, the very young shoots and leaves are free of the burning juice and can be eaten. Nevertheless, it is always wise to wear gloves when dealing with nettles, because the old-timers among them are quick to defend their young.

The tiny green and insignificant flowers that appear in branched clusters or in loose racemes in the leaf axils are the source of the nettle's specific name, *dioica*, meaning "two houses." That is, the flowers are either male or female, and their respective stamens and pistils do not cohabitate as is customary with other flowers. Consequently, the nettle's arrangement of what might be described as casual sex allows the flowers to rely entirely on wind pollination.

Nettles thrive almost anywhere throughout the temperate regions of the world, in waste places and barnyards, along roadsides and hedgerows, quite often in gardens and in moist places. Most of all, nettles thrive in places where the soil is rich in nitrogen.

The process of weaving linen was known as early as the Late Bronze Age, but it was linen made from stinging nettles, whereas flax was still largely regarded as a grain food. It is hardly surprising, therefore, that the preparation and use of nettle fiber should become more refined in the course of succeeding ages. In time, nettles were not only popularly cultivated as a garden crop for their many uses but even tithed.

Nettle cloth could be and was made into any desired texture, from the finest linen for clothing, napery, and bed linen, to the coarsest, such as sacking, sailcloth, and cordage. In the early nineteenth century, the yarn spun from nettles was particularly recommended for making the twine for fishing nets. And a man who is never more specifically identified than as "the poet Campbell" (Thomas?, 1777–1844, or William Wilfred?, 1858–1918) tells us that "in Scotland I have eaten nettles, I have slept in nettle sheets, and dined off a nettle tablecloth. The young and tender nettle is an excellent potherb. The stalks of the old nettle are as good as flax for making cloth. I have heard my mother say that she thought nettle cloth more durable than any other species of linen."

Toward the end of the nineteenth century, the use of flax and hemp superseded that of nettles, not because the latter was considered inferior but because its need to grow in rich soil led to higher costs. Nevertheless, when German and Austrian cotton supplies ran short during World War I, the stinging nettle experienced a resurgent popularity. Experiments showed that the nettle alone met all the conditions of a satisfactory source of textile fiber. At first mixed with 10 percent cotton, nettle fiber was manufactured into such varied goods as underwear, stockings, cloth, and tarpaulins, as well as army clothing. To meet the heavy demands, Austria cultivated nettles on a national scale, and cotton was replaced by a fiber extracted from a tropical relative of the nettle, itself valuable in the production of gas masks.

In 1916, Germans collected slightly under 6 million pounds of nettle fiber for the manufacture of army clothing—more than double the previous year's harvest. By 1918, German army orders unequivocally described nettle as the only efficient cotton substitute. Like cotton, it could be bleached and dyed; mercerized, it was only slightly inferior to silk; conversely, it was considered greatly superior to cotton for velvet and plush.

The huge production of nettle by-products included substitutes for sugar, starch, protein, and ethyl alcohol. Nettle fibers were also used in the manufacture of paper. Not least, German military instructions commanded the inclusion of dried or wilted nettle leaves in the rations of undernourished cavalry horses, because of the plant's high content of fat and protein. And even now, nettles are a commercial source of chlorophyll.

The use of nettles as fodder during World War I was not nearly as innovative as it may seem. Long before then, many horse traders of yore, a breed not generally known as slow to recognize a whinnying edge, are said to have mixed nettle seeds with oats and other feed to sleeken the coats of their merchandise. Although no sensible quadruped would touch a fresh nettle, farmers in Scandinavia, Russia, and Europe learned centuries ago that cows not only relish the taste of dried nettles in their hay but reward such generous donations with increased yields of milk. Farmers also knew that nettle seeds and the dried, powdered greens of the weed mixed with grains kept their poultry healthy and fattened, and increased the production of eggs. Pigs, too, like nettles—as long as the weed is served boiled. They're nobody's fools.

Even the birds and bees benefit from nettles. It seems that the plant's lime and sodium content helps to keep our avian friends healthy, and a stand of nettles is said to protect beehives from the predatory advances of frogs. And lastly, although stinging nettles are a deterrent to flies—a fresh bunch hung in the kitchen keeps it free of them—nettles are host to numerous insects that keep garden pests under control. Some people firmly believe that a planting of stinging nettles will also stimulate plant growth in their vegetable beds and increase the essential oils in their herbs.

A strong decoction made from nettles was already known to Russians of earlier centuries as a means of producing a rich green permanent dye for their woolen clothing and blankets. From nettle roots boiled with alum, they obtained a deep yellow dye with which they colored yarns as well as their Easter eggs.

Not least, a decoction made from boiling stinging nettles in a strong saline solution will curdle milk, thus providing a substitute for rennet. And that brings us to the first—namely, culinary—of the three major remaining uses for which stinging nettles have been known through the ages, the other two being medicinal and cosmetic.

Unlikely a candidate as the stinging nettle may seem to be for culinary purposes, its history as such is nevertheless age old. Even today, nettles are consumed as food or drink in several forms. To a greater or lesser degree, they are a valuable source of vitamins C and D, of carotene, iron, and calcium, as well as numerous other minerals.

One of my own early recollections is of Mother sending my sister, Helen, and me at least once a week in spring to pick young nettle shoots and leaves. We usually collected our harvests along a nearby brook, or from beside a large cow barn that pampered the nettles with all the fertilizer they could have dreamed of.

To this day, I remember the delicious, slightly acrid, salty flavor of the soup Mother made with the greens, or when she added the leaves to salads, steamed them as a vitamin-rich vegetable, or when she baked them in a nourishing cas- serole with potatoes and onions. We never wore gloves when harvesting; we were far too cocky to do so.

Nettles should always be picked only when their shoots are new and tender, no taller than about 8 inches, or while the plant's upper leaves are still pale green. By gathering them regularly, new growth is stimulated well into early summer. Before being cooked, the tops and leaves should always be rinsed under cold running water, swished about with a wooden spoon. Lifted with kitchen tongs, still dripping wet, into a saucepan, the greens are covered and cooked, without added water, for 3 to 5 minutes.

For a simple vegetable, the nettles are then merely strained (I save this cooking water for soup), seasoned with pepper and salt, and served. For an equally nutritious, slightly more elaborate green, I sauté sliced onions until golden, then add the whole fresh nettle tops and leaves, season them with pepper and salt, and add a sprinkling of hot paprika. Then I cook them over medium heat for 5 to 8 minutes, stirring occasionally, until the nettle juice is absorbed, then serve. Mother also used to add the *tenderest* nettle tops to mixed green salads, for a taste not unlike chicory.

For a quick, satisfying soup, I thicken the above-mentioned strained nettle water with dehydrated potatoes, season, and serve it with a sprinkling of grated cheese and fresh, chopped chives. When I have more time, I start the soup with a roux: 1 tablespoon flour is lightly browned in an equal amount of melted butter, in which an onion has been sautéed; the mixture is then thinned with water to the consistency of canned tomato soup, and simmered for 10 minutes. To this I add finely chopped fresh nettles, let the soup cook for another 10 minutes, or until the nettles are tender, and adjust the seasoning.

Mother also used to make a kind of nettle pancake. She first boiled the nettle tops and leaves in a little salted water and then thoroughly drained, strained, and chopped them. Next, she mixed them with grated raw or boiled potatoes, lightly stirred in a beaten egg, some pepper, salt, and chopped fresh chives, shaped the mixture into roughly 4-inch pancakes, and fried them in a little butter, margarine, or olive oil, to be eaten hot or cold. They made favorite school lunches for us.

The kind of rib-sticking casserole Mother used to make for us during wartime food shortages was a satisfying meal for two people: Layer 3 diced raw potatoes with 2 sliced onions, 3 sliced carrots, and 1 quart or more chopped nettles. Seasoned with pepper and salt, plus a light dusting of ground caraway seed, the casserole is baked, covered, at 350 degrees Fahrenheit, ready to eat in 45 to 50 minutes—delicious and colorful.

Nettle pudding was long a favored dish in Scotland and England, and is recommended even today among people who value cooking with wild plants. To prepare this recipe for six persons: 1 pound washed, chopped young nettle tops, 2 heads broccoli (or 1 small cabbage, or some brussels sprouts), and 2 large leeks, together with ¼ pound rice, are placed in a muslin bag and boiled in salted water to cover, until the vegetables are tender. It is served hot with a rich brown gravy or with butter browned to a nutty flavor.

A diary entry of Samuel Pepys, dated February 1661, notes that when he called on a Mr. Simons in London, that gentleman was abroad, but his wife, "like a good lady, within, and there we did eat some nettle porridge, which was made on purpose to-day for some of their coming, and was very good." The ingredients of a similar dish made with fresh greens by some modern naturalists include 6 ounces nettles, 1 ounce dandelions, 1 ounce sorrel, 2 ounces watercress, and 1 onion; the added flavorings, apart from pepper and salt, are mint, black currant leaves, and thyme, and the thickening agent is 1 beaten egg. Washed and chopped, the greens are mixed with all other ingredients and steamed 1½ hours in a covered, heatproof bowl set in a water bath in the oven (about 350 degrees Fahrenheit), or on the stove over medium heat.

Black currant leaves are an optional addition in the making of nettle beer. This is not only a pleasant, refreshing drink but is quite popular among older country folk in England, as a remedy for rheumatic pains—or so they say! To a large enamel saucepan they add 2 gallons cold water, a tightly packed pailful of washed nettle tops, 4 handfuls each dandelions and cleavers—or juice of 2 lemons—and 2 ounces bruised fresh whole ginger. This they boil gently for 40 minutes, strain, then stir in 2 cups brown sugar, and cool to lukewarm. Laid atop the liquid is one slice of toast spread with a paste made of 1 ounce compressed fresh yeast mixed with 1 teaspoon sugar. The pot is then covered with a lint-free towel and kept warm for 7 hours, or overnight. Once the scum is removed and 1 tablespoon cream of tartar is stirred into the liquid, the new beer can be bottled.

I have only one regret nowadays, and that is the lack of stinging nettles growing anywhere near me. It is surely easy to see, however, how the stinging nettle came to be deliberately introduced into America by early immigrants.

In spite of the many medicinal properties attributed to the stinging nettle in the course of its history, modern medicine has no particular use for the plant. Nevertheless, in early folk medicine, asthmatics are said to have burned dried nettle leaves and inhaled the smoke for relief. John Gerard agreed that the weed was beneficial for this affliction when he wrote, "The nettle is a good medicine for them that cannot breathe unless they hold their heads upright." Nicholas Culpeper added numerous other recommendations. "The roots or leaves boiled," he wrote, "or the juice of either of them, or both, made into an electuary [confection] with honey and sugar, is a safe and sure medicine to open the passages of the lungs . . . it likewise helps the swelling of both the mouth and throat if they be gargled with it." A decoction of leaves or seeds, he claimed, "provokes urine, and expels gravel and stone," whereas the expressed nettle juice, or a "decoction of the roots, is good to wash either rotten, or stinking sores or fistulas, and gangrenes, and such as fretting, eating, or corroding scabs, manginess, and itch in any part of the body."

Long before both these herbalists, Galen, recognizing other virtues in nettles, had written in the second century A.D. that the seeds "taken in a draught of mulled wine, they arouse desire." Paracelsus, the great physician and herbal healer, recommended a twice daily drink of expressed nettle juice, made from the roots, mixed with goat's milk.

Nettles have been used for diarrhea, enteritis, and gout, and to treat edema, arthritis, stomach cramps, bed-wetting, and impotence. They have been employed to stop nosebleeds and to control hemorrhages and hemophilia. Swiss homeopath Dr. Alfred Vogel asserts that "no other plant can equal the nettle in cases of anemia, chlorosis [greensickness], rachitis (rickets), scrofula, respiratory diseases, and especially lymphatic troubles." He also notes that a drink made of young spring nettles boiled in milk not only relieves constipation but also stops migraines.

Modern American as well as European homeopaths concur in the traditional belief in nettle as a blood purifier, as antiscorbutic and astringent. They not only prescribe nettle tea for asthma and anemia but recommend it as an invigorating, purgative tonic because of the weed's high vitamin C content. They consider nettle juice mixed with honey a valuable treatment for chest and lung troubles, and they place their trust in the external use of nettle powder to stanch bleeding. Herbalists have also found that two teacups of nettle tea taken daily are helpful in cases of blood pressure disorders. Even cases of iron deficiency that have failed to respond to advanced drugs are said to have yielded to the powers of the nettle. Compresses dipped in a nettle infusion are applied to burns and cuts, on the same principle as the long popular application of nettles as a counterirritant: By being placed on inflamed skin, nettle acts as an irritant, thus stimulating increased bloodflow to the affected area, which in turn reduces the inflammation.

In Germany, the nettle has long been a favorite treatment for neuralgia—4 tablespoons of a nettle decoction being taken three times daily, or a poultice of fresh, bruised leaves being applied to the affected area. In India, a decoction of nettle is prescribed for kidney disorders and tuberculosis, as well as for hemorrhages of the kidneys and uterus. In Russia, the plant is used for the relief of toothache and sciatica. The Romans steeped the fresh leaves in olive oil to prepare a healing salve. Jamaicans drop the fresh juice into open wounds.

Among Germans and Russians particularly, urtication was a respected remedy for chronic rheumatism, in which the afflicted part of the body was at least twice daily rubbed or flogged with a bunch of fresh nettles. The priestly herbalist Father Kneipp recommended this practice in rheumatic cases that had failed to respond to all other treatments. Although Kneipp does not indicate whether or not he spoke from personal experience, he said that "the fear of the unaccustomed rod will soon give way to joy at its remarkable healing efficacy." Understatement though they are, these may well have been the very words spoken by libertines of ancient Rome, after urtication had rekindled their exhausted sexual appetites. That is, if the accounts of Petronius, Nero's Arbiter of Taste, are to be trusted!

Cosmetically, there is little danger that nettles will ever invade the mirrored meccas of makeup mavens. But nettles do enjoy a long reputation as a rubefacient and as a most refreshing tonic for face, body, scalp, and hair. For this 1 cup boiling water is poured over 1 teaspoon dried herb, steeped for 30 minutes until cool, then strained and used. This wash can also be liberally splashed on the skin when one is tired, or used as a rinse for hair after shampooing, or as a daily tonic rubbed into the scalp to stimulate hair growth. For this last purpose, a modern German physician recommends rubbing the following tonic into the scalp: To finely chopped fresh nettles an equal amount of half water, half cider vinegar is added, brought to a boil and simmered 15 to 20 minutes, strained, and bottled. Herbalists also use nettle infusions for controlling dandruff.

A strong infusion of fresh nettle, together with comfrey leaves and stems added to the bath, is used to stimulate blood circulation, act as tonic for the entire body, and to soothe the skin. For a skin-freshening cleansing oil, 3 tablespoons bruised fresh nettle leaves and 3 ounces sunflower, safflower, or almond oil are placed into a small bottle, and allowed to macerate for two days, and frequently shaken during this period.

Preparatory to a facial steam and face pack, I cleanse my skin with the oil, then sit over a nettle steam (1 quart boiling water poured over 4 tablespoons bruised fresh herb), with a towel placed over my head so that the steam cannot escape. At the end of 15 to 20 minutes I gently dab my face with a cool, damp cloth.

For the face pack that follows, I mix 2 tablespoons uncooked oatmeal with enough milk to make a thick paste, then thin this just enough with fresh nettle juice so that the paste can be easily spread. I leave it to dry for 15 minutes before rinsing the skin with lukewarm water, patting it dry, and applying a moisturizer. Sheer rejuvenation and bliss is this indulgence.

Not to be forgotten, for anybody who is subject to chilblains, is the fact that nettles have been used to treat them for centuries. I learned as a child that the raw juice, or a strong infusion cooled, when applied on a compress to the afflicted area acts as a counterirritant. In doing so, the nettle infusion reduces the burning sensation of the chilblains, and helps the area to recover.

As though its history alone were not enough, various magical powers have also been ascribed to the nettle. An overweight diabetic, for instance, is said to have reported amazing results from an astonishing treatment of his disease in 1926. Following a two-day fast, he claimed, a diet of nothing but young nettles and nettle tea not only improved his condition but also reduced his weight by 78 pounds! Similarly, three days after a woman afflicted by severe rheumatism had accidentally brushed against nettles while blackberry picking, she was reported as being virtually free of her disease—and as *remaining* free of it. In Austria, Tyroleans believed that throwing fresh nettles into the fire during a storm would protect them from lightning. And country people in France believed that a bunch of nettles and yarrow held in the hand would quell a person's fear. Well, of *course*—the searing burn from the nettles would by far transcend all mortal fear!

Verbascum thapsus

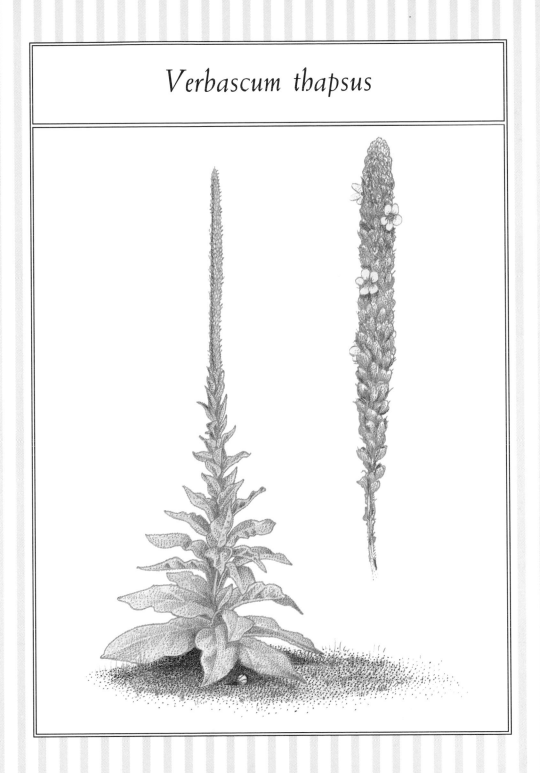

Verbascum thapsus

(GREAT MULLEIN)

VITAL STATISTICS

Biennial

COMMON NAMES	Great Mullein, Common Mullein, Aaron's Rod, Jacob's Staff, Torches, Our Lady's Flannel, Velvet Dock, Blanket Herb, Candlewicks, Clown's Lungwort, Flannel Leaf, Hedge Taper, Beggar's Blanket, Rag Paper, Wild Ice Leaf, Jupiter's Staff, Shepherd's Staff, Golden Rod, Adam's Flannel, Feltwort, Fluffweed, Hare's Beard, Hag's Taper, Torchwort, Peter's Staff, Agleaf, Bunny's Ears
USES	Medicinal, cosmetic
PARTS USED	Leaves, flowers, roots
HEIGHT	To 8 feet, mostly 4 to 6 feet, unbranched stems
FLOWER	Yellow, five petaled, on long spike, capable of self-fertilizing; June to September
LEAVES	6 to 15 inches long, broad, ovate, grayish white, woolly, alternating
ROOT TYPE	Short taproot, easily extracted
HABITAT	Fields, roadsides, wild gardens, poor soil
PROPAGATION	Seeds
CONTROL	Weeding

Mullein is surely one of the noblest weeds. It instantly commands our attention, whether it stands like a solitary beacon in a field, or by the roadside, or in an abandoned building lot. At times, it even appears in our gardens, from a seed blown there on a whim of nature.

In the first year of its growth, only a rosette of its furry leaves appears, somewhat similar to the foxglove, except that mullein leaves are thicker, hairier, and more silvery. The stem emerges in the second year. It grows slender, upright, and rigid, to an average height of 4 to 6 feet, tapering into a narrow spire, the 1-foot flower spike, whose crowded buds open randomly into stalkless blossoms of purest gold from June to September.

Arranged on alternate sides of the stem are the silvery gray, broad, woolly leaves, in ever decreasing size, with the base of each leaf extending some distance down the stem. Because mullein grows mostly in poor, dry soils, this leaf system allows water to be directed from the smaller leaves above to the larger below, and from them to the roots. The thick covering of hairs on the leaves not only protects them against loss of moisture but also against insects. And, because the leaves provoke acute irritation in the mucous membranes, browsing animals, domestic and wild, give the plant wide berth. Even grass is unable to crowd mullein, which makes the weed all the easier to eradicate.

Linnaeus named it *Verbascum*, which is believed to be a corruption of *barbascum*, from the Latin *barba*, for "beard," perhaps in allusion to the plant's hairy foliage. The specific name *thapsus* may refer to the North African town of Caesar's victory, from which Caesar, conceivably, may have introduced mullein to Rome; or else it derives from *Thapsia*, a genus of European plants resembling mullein. The common name is thought to have originated with *malandria*, Latin for "blisters or sores," from which derives the English *malanders* (actually an affliction of horses, but also used as meaning leprosy); others suggest that the root of the word is to be found in the Latin *mollis*, for "soft," in allusion to the plant's soft leaves.

Although the exact origins of both the plant and its names have been blurred by passing time, mullein (pronounced "mullen") is a native from the Mediterranean Basin and spread throughout Europe and North Africa, as well as in temperate west and central Asia and North America. It has been known and used since time immemorial, invested with the powers of magic, myth, and majesty. It is the plant on whose power Ulysses is said to have relied to protect him against the wiles of Circe. In India and eastern Europe, mullein was believed to have the power to drive away evil spirits.

"Verbascum," John Parkinson wrote in the sixteenth century, "is called of the Latines Candela regia [king's candle], and Candelaria . . . to burne, at funeralls or otherwise." This was accomplished easily enough. First stripped of leaves, then dipped in suet, pitch, or resin, and lighted, mullein stalks served as the candles and torches of ancient kings and commoners alike. Until cotton came into common use, the down of the leaves served for tinder and lamp wicks, because it ignited upon even the slightest spark.

Beggars throughout history are said to have found warmth under the large woolly leaves of the plant; one region believed that mullein leaves placed in shoes ensured conception, whereas another region believed exactly the opposite. One thing is certain: Woolly mullein leaves slipped into shoes have warmed many a poorly shod foot throughout the ages, including in modern times of great need.

In remote areas of Austria's Tyrol, mullein blossoms have been popularly credited with accurately predicting the weather. If the blooms are low on the stalk, an early snow can be expected; a long winter with heavy snow late in the season should be reckoned with if the blossoms are at the end of the stalk. In Germany's Rhineland, the customary annual consecration of herbs has for centuries included mullein. In southern Germany, some Catholics still place mullein in the center of an herbal bouquet to be blessed in a special religious service on Assumption Day. The bouquet is then carefully preserved against possible disasters in the coming months: Should there be illness among the livestock, some dried flowers are added to the fodder to restore the animals' health; should severe storms threaten to damage the home, the flowers are thrown into the hearth fire for protection.

Not least, some German fishermen even today assure themselves of a plentiful catch, as did their forebears, by first strewing mullein seeds into the waters they plan to fish the next day. By doing this, they follow an ancient ritual that can be traced back more than 2,000 years. And seemingly it works because the mild narcotic properties contained in the seeds are said to intoxicate and numb the fish!

But by far the plant's most consistent application, throughout its long history, has been in matters medicinal. Dioscorides and Galen are recorded to have treated pulmonary illnesses with the plant's roots; Pliny prescribed mullein leaves for bronchial ailments in humans and horses. Saint Hildegard, the medieval German abbess and mystic, regarded both the plant's flowers and leaves as a specific for hoarseness.

Some medieval physicians went to considerable lengths to prepare what they must have considered a surefire cure for gout and hemorrhoids. And perhaps it was. After pounding together the flowers and leaves of mullein in a wooden tub, they sealed the tub with plaster and set it in the sun, or buried it in dung for three months. When they deemed the "digestion" to be completed, they pressed the juice from the mullein's remains and stored it in tightly corked jars, ready for external application. Some centuries later, Culpeper had greatly simplified remedies for both afflictions. "The oil made by infusion of the flowers," he wrote, "is of good effect for the piles. . . . Three ounces of the distilled water of the flowers drank morning and evening is a remedy for the gout."

As we have come to expect of him, Culpeper, of course, also attributed numerous other remedies to mullein. For example, he suggested small quantities of the root "against laxes and fluxes of the body. The decoction, if drunk, is profitable for . . . cramps, convulsions, and old coughs. The decoction gargled, eases tooth-ache." Warts could be removed, he claimed, by rubbing on them either the juice of the flowers and leaves, or the powder of dried mullein roots. A decoction made of the leaves and roots was greatly effective, he added, in dissolving tumors, swellings, and inflammations of the throat. The seeds boiled in wine, together with the leaves, not only speedily extracted "thorns or splinters from the flesh," but also helped the skin to heal. The seeds alone, "bruised and boiled in wine," eased the pain and swelling of "any member that has been out of joint, and newly set again."

Among the American Indians, the Navahos are known to have long ago named the plant "big tobacco." Not only did they, like other tribes, smoke it in their pipes to relieve a sore throat but, in order to ease mental distress, they are reported to have smoked mullein mixed with real tobacco. A Dr. A. W. Chase, who is not otherwise identified, related that "an old gentleman, an inveterate smoker, from my suggestion, began to mix the mullein with his tobacco, one-fourth at first, for awhile; then half, and finally three-fourths; at this point he rested. It satisfied in place of the full amount of tobacco, and healed a cough which had been left upon him after inflammation of the lungs." Similarly, a nineteenth-century journal, *Medical and Surgical Reporter*, stated that "the mullein smoked through a pipe acts like a charm and affords instant relief."

From the standpoint of modern herbal medicine, it might almost be said that the redoubtable Culpeper actually overlooked some uses of this weed. For instance, mullein is described as having remarkable demulcent, emollient, and anti-spasmodic properties; as being a valued astringent, diuretic, vulnerary, and expectorant, rich in mucilage and other valuable constituents. And it is considered particularly useful in treating complaints of the respiratory system, wasting diseases, and inflammations of the mucous membrane, and as a respected nerve tonic.

The radiant yellow blossoms with their faint honey scent are used both fresh and dried. They are usually plucked as soon as they open—and only on dry, sunny days. If they are to be stored, they can be air dried for two or three days, then placed on a baking sheet covered with brown paper, and completely dried in the oven at a setting no higher than 180 degrees Fahrenheit before being stored in tightly closed containers. The containers, preferably glass, should be kept in a dark, dry location, away from direct light.

The leaves are easy to dry, although it takes longer. I generally place them in the attic or basement, both of which are dry, or else in a dry and airy spare room. Spread on a double thickness of newspaper, they may need three weeks before being ready for storage. As with all large-leaved herbs, it is best not to break or crumble the leaves until they are about to be used, as this best preserves their beneficial properties.

What cannot be stressed enough, whether any part of mullein is to be used fresh or dried in an infusion, decoction, or oil, is the need for straining these liquids through a double layer of muslin, in order to eliminate the fine hairs that cover all parts of the plant, including the flowers. These hairs can otherwise cause itching in the mouth, even after they have been steeped in liquid.

My own introduction to mullein's curative properties came in the early

1960s. Subject as I was to a recurrent bronchial affliction, I needed little persuasion to try the recipe Euell Gibbons included in his book *Stalking the Wild Asparagus*. I have made it a practice ever since to keep a bottle of this cough syrup on hand for the winter months. The ideal time to prepare the mixture is July, when all four ingredients are available and at their best.

I gather 1 cup each of red clover blossoms (*Trifolium pratense*), chopped new needle growth of white pine (*Pinus strobus*), and finely chopped mullein leaves, plus ½ cup of finely chopped *inner* bark from any of the wild cherries. In a saucepan, I cover all four ingredients with 1 quart water, bring to a quick boil, immediately lower the heat, and gently simmer it for 20 to 30 minutes. I strain the juice through a double layer of muslin (to eliminate the hairs of the mullein leaves), and add 1 pint of pure honey to the juice, then bring this mixture to a rapid boil, pour it into tightly capped bottles, and store in the refrigerator. By taking 1 to 2 teaspoonfuls of the syrup every four hours, I have rarely had need to take it for a second day.

Perhaps by far the most common method of preparing mullein is in an infusion, using 1 ounce of dried or 2 ounces of fresh leaves to a pint of water. With or without honey added, not only is this a pleasant tea but, to be taken in frequent doses of a wineglassful, herbalists prescribe it for relieving a dry cough, allaying diarrhea, and acting as a mild nerve tonic.

For a decoction (1 ounce of dried or 2 ounces of fresh leaves) the herb is brought to a boil in 1 pint of milk, simmered gently for 10 minutes, and strained. No more than three wineglassfuls are recommended, taken warm, sweetened with honey or not. It is considered both a pleasantly soothing and a nutritious medicine, especially for those suffering from wasting diseases or severe coughs. And even the decocted leaves still have one more popular purpose in them, should there be need of it: Applied as a poultice, they can bring immediate relief from the ache of boils, ulcers, frostbite, piles, scabs, and burns.

A sweetened infusion made of the flowers and water has been a long-trusted domestic remedy both in England and on the Continent. There it is prepared for colds, coughs, colic, and catarrhs, as well as for sore throats. The renowned late nineteenth-century Bavarian priest Sebastian Kneipp asserted that snuffing such a tea up a congested nose instantly clears it. Certainly this echoes John Gerard's similar claim that the juice of mullein root, snuffed up the nose, was a remedy for migraine. Kneipp also recommended mullein blossoms for their restorative action on the heart, particularly if a few flowers were routinely added to other soup greens in the preparation of a strong beef broth.

However, perhaps the most favored preparation made from the plant is mullein oil. Such an oil is considered one of the finest remedies for all ear complaints, a few drops applied overnight reportedly effecting a complete cure. In certain rural areas of Germany even today, mullein oil, as an alternative to teas or compresses, is kept on hand not only to treat many of the above-named afflictions, but also because it is regarded as a valuable bactericide. And "because of its reliability and complete efficacy," Father Kneipp considered mullein oil "as good as a specific" for incontinence of urine. He wrote of a sixteen-year-old boy, a lifelong sufferer of this disorder, who was restored to complete health after daily taking 15 drops of mullein oil for only a brief period.

Mullein oil is produced by macerating one part fresh flowers in two parts of olive oil. The mixture is placed in a corked or screw-top bottle or jar, and set in direct sunlight for about three weeks, the bottle shaken once or twice daily. At the end of this period, the oil is strained, bottled, and stored for use. Or, in cool or overcast weather, the bottle can be placed on the back of a stove, near a fire, or on a warm radiator. For yet another method, the blossoms and oil can be steeped in a saucepan for an hour, over *very* low heat, then strained and stored.

Finally, the women of ancient times obviously learned something about mullein that even learned men might not otherwise have discovered about the plant. According to the writings of both Dioscorides and Galen, as well as of later writers, it seems that the yellow dye the women obtained from infusing the blossoms served them as a popular rinse that added golden highlights to their hair. There is also no reason to doubt that the same women passed down to us, through the ages, yet another cosmetic use of the plant. In the New World, this use seems to have made its way into a local newspaper in the early part of this century. It appears that girls in Wisconsin often rubbed their cheeks with mullein leaves to make them rosy. Certainly, this makes good sense. After all, the leaves worn in shoes or rubbed on cold skin have a long tradition of stimulating blood circulation.

Mullein may be only a weed, but with so much history, tradition, and lore, surely it has earned a place not only in our medicine cabinets but also in our lives and our garden beds. Perhaps it would gain such popular acceptance, as it has in European gardens, if our common name for it suggested the plant's inherent nobility.

APPENDIX A: TRADITIONAL HERBAL FOLK REMEDIES

Abortive
Achillea millefolium

Abscesses *see* BOILS/ABSCESSES/CARBUNCLES

Acne
Agropyron repens
Arctium lappa
Daucus carota
Tussilago farfara

Allergies, Skin
Daucus carota

Anemia
Achillea millefolium
Cichorium intybus
Equisetum arvense
Nasturtium officinale
Plantago major
Rumex crispus
Urtica dioica

Antibiotics
Equisetum arvense

Anticonvulsant *see* EPILEPSY

Antiperspirant *see* DEODORANT

Antiscorbutics
Capsella bursa-pastoris
Chenopodium album
Daucus carota
Glechoma hederacea
Nasturtium officinale
Rumex acetosella
Taraxacum officinale
Urtica dioica

Antiseptics
Achillea millefolium
Berberis vulgaris
Daucus carota
Matricaria recutita
Rhus glabra

Antispasmodics
Achillea millefolium
Chrysanthemum leucanthemum
Matricaria recutita
Rumex acetosella
Verbascum thapsus

Aperients *see* LAXATIVES

Aphrodisiacs
Arctium lappa
Nasturtium officinale
Urtica dioica

Arteriosclerosis
Capsella bursa-pastoris
Taraxacum officinale

Arthritis/Rheumatism/Gout
Achillea millefolium
Agropyron repens
Arctium lappa
Berberis vulgaris
Brassica nigra
Capsella bursa-pastoris
Chelidonium majus
Chrysanthemum leucanthemum
Cichorium intybus
Daucus carota
Equisetum arvense
Matricaria recutita
Melilotus officinalis
Nasturtium officinale
Phytolacca americana
Rhus radicans
Rumex crispus
Senecio vulgaris
Stellaria media
Taraxacum officinale
Urtica dioica
Verbascum thapsus

Asthma
Daucus carota
Glechoma hederacea
Tussilago farfara
Urtica dioica

Astringent
Achillea millefolium
Berberis vulgaris
Capsella bursa-pastoris
Equisetum arvense
Matricaria recutita
Plantago major
Rhus glabra
Rumex crispus
Urtica dioica
Verbascum thapsus

Bactericide
Verbascum thapsus

Bed-Wetting *see* URINE INCONTINENCE

Bile Disorders *see* GALLBLADDER/BILE DISORDERS

Bladder Disorders *see* URINARY DISORDERS

Bleeding/Hemorrhages
Achillea millefolium
Capsella bursa-pastoris
Equisetum arvense
Galium aparine
Plantago major
Rumex acetosella
Senecio vulgaris
Urtica dioica

Blisters
Chelidonium majus
Galium aparine

Blood Pressure
Berberis vulgaris
Capsella bursa-pastoris

Taraxacum officinale
Urtica dioica

Boils/Abscesses/Carbuncles
Arctium lappa
Daucus carota
Equisetum arvense
Glechoma hederacea
Matricaria recutita
Phytolacca americana
Rumex acetosella
Senecio vulgaris
Stellaria media
Tussilago farfara
Verbascum thapsus

Bone Injuries
Equisetum arvense

Bronchitis *see* UPPER RESPIRATORY INFECTIONS

Bruises
Arctium lappa
Chrysanthemum leucanthemum
Plantago major

Burns/Scalds
Arctium lappa
Daucus carota
Galium aparine
Plantago major
Rumex crispus
Urtica dioica
Verbascum thapsus

Carbuncles *see* BOILS/ABSCESSES/CARBUNCLES

Cataracts
Chelidonium majus

Catarrhal Conditions
Achillea millefolium
Agropyron repens
Capsella bursa-pastoris
Chelidonium majus
Glechoma hederacea
Matricaria recutita
Nasturtium officinale
Phytolacca americana
Taraxacum officinale
Tussilago farfara
Verbascum thapsus

Chapped Skin
Glechoma hederacea
Senecio vulgaris

Chilblains/Frostbite
Brassica nigra
Daucus carota
Urtica dioica
Verbascum thapsus

Chlorosis
Urtica dioica

Cholesterol
Taraxacum officinale

Circulation Disorders
Achillea millefolium
Berberis vulgaris
Capsella bursa-pastoris
Chelidonium majus

Taraxacum officinale
Urtica dioica

Colds *see* UPPER RESPIRATORY INFECTIONS

Colic
Matricaria recutita
Senecio vulgaris
Verbascum thapsus

Compresses
Equisetum arvense
Rumex acetosella
Tussilago farfara
Urtica dioica

Constipation *see* LAXATIVES

Corns
Chelidonium majus

Coughs *see* UPPER RESPIRATORY INFECTIONS

Counterirritant
Rhus radicans
Urtica dioica
Verbascum thapsus

Deodorant/Antiperspirant
Equisetum arvense
Galium aparine
Nasturtium officinale

Depressants
Berberis vulgaris

Depression
Achillea millefolium
Melilotus officinalis

Depuratives
Agropyron repens
Arctium lappa
Cichorium intybus
Galium aparine
Glechoma hederacea
Phytolacca americana
Plantago major
Rumex crispus
Stellaria media

Diabetes
Nasturtium officinale
Taraxacum officinale

Diaphoretics
Achillea millefolium
Agropyron repens
Arctium lappa
Matricaria recutita
Rumex acetosella
Verbascum thapsus

Diarrhea
Berberis vulgaris
Capsella bursa-pastoris
Daucus carota
Matricaria recutita
Plantago major
Rhus glabra
Rumex crispus
Taraxacum officinale
Urtica dioica
Verbascum thapsus

Digestive Aids
Achillea millefolium

Arctium lappa
Berberis vulgaris
Brassica nigra
Capsella bursa-pastoris
Cichorium intybus
Matricaria recutita
Melilotus officinalis
Nasturtium officinale
Rumex acetosella
Rumex crispus
Taraxacum officinale

Diuretics
Achillea millefolium
Agropyron repens
Brassica nigra
Capsella bursa-pastoris
Chenopodium album
Chrysanthemum leucanthemum
Daucus carota
Equisetum arvense
Galium aparine
Matricaria recutita
Melilotus officinalis
Plantago major
Rhus glabra
Rumex acetosella
Rumex crispus
Stellaria media
Verbascum thapsus

Dizziness
Achillea millefolium
Matricaria recutita

Dropsy *see* EDEMA

Dysentery *see* DIARRHEA

Earache
Achillea millefolium
Capsella bursa-pastoris
Galium aparine
Matricaria recutita
Melilotus officinalis
Plantago major
Rumex crispus
Verbascum thapsus

Eczema
Arctium lappa
Chelidonium majus
Equisetum arvense
Nasturtium officinale
Rumex crispus
Taraxacum officinale

Edema
Berberis vulgaris
Cichorium intybus
Daucus carota
Equisetum arvense
Galium aparine
Matricaria recutita
Taraxacum officinale
Urtica dioica

Emollients
Stellaria media
Verbascum thapsus

Enema
Tussilago farfara

Enteritis *see* GASTROENTERITIS

Epilepsy
   Berberis vulgaris
   Senecio vulgaris

Eructation *see* FLATULENCE/ERUCTATION (GAS-X)

Erysipelas (St. Anthony's fire)
   Capsella bursa-pastoris

Expectorant
   Tussilago farfara
   Verbascum thapsus

Eyes, Eyelids
   Achillea millefolium
   Arctium lappa
   Berberis vulgaris
   Chelidonium majus
   Chrysanthemum leucanthemum
   Cichorium intybus
   Equisetum arvense
   Glechoma hederacea
   Matricaria recutita
   Melilotus officinalis
   Phytolacca americana
   Portulaca oleracea
   Rumex crispus
   Senecio vulgaris
   Stellaria media
   Taraxacum officinale
   Tussilago farfara

Eyesight/Night Blindness
   Chelidonium majus
   Daucus carota
   Nasturtium officinale

Facial Blemishes/Freckles
   Achillea millefolium
   Brassica nigra
   Cichorium intybus
   Equisetum arvense
   Galium aparine
   Taraxacum officinale
   Tussilago farfara

Fevers
   Achillea millefolium
   Arctium lappa
   Berberis vulgaris
   Cichorium intybus
   Equisetum arvense
   Matricaria recutita
   Nasturtium officinale
   Plantago major
   Portulaca oleracea
   Rhus glabra
   Rumex acetosella
   Stellaria media

Flatulence/Eructation (Gas-X)
   Achillea millefolium
   Daucus carota
   Matricaria recutita
   Melilotus officinalis
   Rumex crispus

Foot Care
   Arctium lappa
   Brassica nigra
   Glechoma hederacea

Freckles *see* FACIAL BLEMISHES/FRECKLES

Frostbite *see* CHILBLAINS/FROSTBITE

Gallbladder/Bile Disorders
Agropyron repens
Berberis vulgaris
Cichorium intybus
Rumex crispus
Senecio vulgaris
Taraxacum officinale

Gallstones
Agropyron repens
Chelidonium majus

Gargles
Berberis vulgaris
Brassica nigra
Equisetum arvense
Rumex acetosella
Senecio vulgaris
Tussilago farfara
Urtica dioica
Verbascum thapsus

Gastric Disorders
Achillea millefolium
Agropyron repens
Capsella bursa-pastoris
Cichorium intybus
Equisetum arvense
Glechoma hederacea
Rumex crispus
Senecio vulgaris
Taraxacum officinale
Tussilago farfara
Urtica dioica

Gastroenteritis
Achillea millefolium

Agropyron repens
Tussilago farfara
Verbascum thapsus

Gum Disorders
Equisetum arvense
Portulaca oleracea

Hair Restorers
Achillea millefolium
Arctium lappa
Brassica nigra
Urtica dioica

Headaches, Migraines
Achillea millefolium
Brassica nigra
Glechoma hederacea
Matricaria recutita
Melilotus officinalis
Phytolacca americana
Portulaca oleracea
Rumex crispus
Urtica dioica
Verbascum thapsus

Hearing
Glechoma hederacea

Heart
Cichorium intybus
Rumex acetosella
Verbascum thapsus

Hemophilia
Urtica dioica

Hemorrhages *see* BLEEDING/HEMORRHAGES

Hemorrhoids
  Achillea millefolium
  Agropyron repens
  Capsella bursa-pastoris
  Chelidonium majus
  Equisetum arvense
  Matricaria recutita
  Phytolacca americana
  Plantago major
  Rumex crispus
  Stellaria media
  Tussilago farfara
  Verbascum thapsus

Hemostatics (stop bleeding) *see* BLEEDING/HEMORRHAGES

Herbal Baths
  Equisetum arvense
  Glechoma hederacea
  Taraxacum officinale

Herbal Tobacco
  Rhus glabra
  Tussilago farfara
  Verbascum thapsus

Herpes
  Plantago major

Hiccups
  Brassica nigra
  Daucus carota

Hypertension *see* BLOOD PRESSURE

Hypoglycemia *see* DIABETES

Impotence
  Urtica dioica

Inflammations
  Achillea millefolium
  Matricaria recutita
  Melilotus officinalis
  Phytolacca americana
  Rumex crispus
  Stellaria media
  Urtica dioica
  Verbascum thapsus

Influenza
  Daucus carota
  Matricaria recutita

Inhalants
  Achillea millefolium
  Glechoma hederacea
  Matricaria recutita

Insect Bites/Stings
  Arctium lappa
  Galium aparine
  Phytolacca americana
  Plantago major
  Rumex crispus
  Senecio vulgaris

Insect Repellent
  Chrysanthemum leucanthemum

Insomnia *see* SEDATIVES

Iron Deficiency
Chenopodium album

Irritants
Brassica nigra
Rhus radicans

Jaundice *see* LIVER DISORDERS

Joints, Painful
Berberis vulgaris
Cichorium intybus
Stellaria media
Taraxacum officinale
Urtica dioica

Kidney Complaints
Achillea millefolium
Agropyron repens
Arctium lappa
Capsella bursa-pastoris
Chelidonium majus
Cichorium intybus
Daucus carota
Equisetum arvense
Glechoma hederacea
Plantago major
Rumex acetosella
Senecio vulgaris
Stellaria media
Taraxacum officinale
Urtica dioica

Laryngitis *see* THROAT INFECTIONS

Laxatives
Brassica nigra

Capsella bursa-pastoris
Chenopodium album
Cichorium intybus
Daucus carota
Galium aparine
Glechoma hederacea
Matricaria recutita
Nasturtium officinale
Phytolacca americana
Plantago major
Rumex acetosella
Rumex crispus
Stellaria media
Taraxacum officinale
Urtica dioica

Lead Poisoning
Glechoma hederacea

Lithontriptics
Agropyron repens
Daucus carota
Equisetum arvense
Galium aparine
Senecio vulgaris

Liver Disorders
Achillea millefolium
Agropyron repens
Berberis vulgaris
Capsella bursa-pastoris
Chelidonium majus
Chrysanthemum leucanthemum
Cichorium intybus
Daucus carota
Equisetum arvense
Glechoma hederacea

Portulaca oleracea
Rumex crispus
Senecio vulgaris
Stellaria media
Taraxacum officinale

Lung Diseases
Achillea millefolium
Capsella bursa-pastoris
Glechoma hederacea
Taraxacum officinale
Tussilago farfara
Urtica dioica
Verbascum thapsus

Lymphatic Problems
Rumex crispus
Urtica dioica

Measles
Achillea millefolium
Arctium lappa

Melancholy *see* DEPRESSION

Menstruation, Menstrual Disorders
Achillea millefolium
Berberis vulgaris
Capsella bursa-pastoris
Equisetum arvense
Matricaria recutita
Plantago major
Senecio vulgaris

Migraines *see* HEADACHES

Mucous Membrane Inflammations
Achillea millefolium
Agropyron repens

Capsella-bursa pastoris
Chrysanthemum leucanthemum
Glechoma hederacea
Matricaria recutita
Tussilago farfara

Mumps
Phytolacca americana

Nails
Equisetum arvense

Nervous Complaints
Achillea millefolium
Equisetum arvense
Matricaria recutita
Nasturtium officinale
Portulaca oleracea
Taraxacum officinale
Verbascum thapsus

Neuralgia
Brassica nigra
Matricaria recutita
Urtica dioica

Night Blindness *see* EYESIGHT/NIGHT BLINDNESS

Nosebleeds
Achillea millefolium
Capsella bursa-pastoris
Equisetum arvense
Rumex crispus
Urtica dioica

Obesity
Chelidonium majus
Taraxacum officinale

Oily Skin
Equisetum arvense

Painkiller
Achillea millefolium

Palsy
Chrysanthemum leucanthemum

Piles *see* HEMORRHOIDS

Poison Ivy
Plantago major
Rhus radicans see JEWELWEED

Poultices
Achillea millefolium
Arctium lappa
Capsella bursa-pastoris
Chenopodium album
Cichorium intybus
Equisetum arvense
Galium aparine
Glechoma hederacea
Matricaria recutita
Plantago major
Rhus glabra
Rumex acetosella
Rumex crispus
Senecio vulgaris
Stellaria media
Tussilago farfara

Psoriasis
Achillea millefolium
Galium aparine

Purgatives
Berberis vulgaris
Rumex crispus
Senecio vulgaris
Urtica dioica

Refrigerants
Arctium lappa
Plantago major
Portulaca oleracea
Rhus glabra
Rumex acetosella
Stellaria media
Tussilago farfara

Relaxants
Arctium lappa

Respiratory-Tract Disorders *see* UPPER RESPI-
RATORY INFECTIONS

Restoratives
Berberis vulgaris
Cichorium intybus
Equisetum arvense
Stellaria media

Rheumatism *see* ARTHRITIS/RHEUMATISM/GOUT

Ringworm
Berberis vulgaris
Chelidonium majus
Plantago major
Rhus glabra
Rhus radicans
Rumex crispus

Rubefacient *see* COUNTERIRRITANT

Scabs
  Arctium lappa
  Chelidonium majus
  Matricaria recutita
  Plantago major
  Rumex acetosella
  Rumex crispus
  Urtica dioica
  Verbascum thapsus

Scalds *see* BURNS/SCALDS

Sciatica
  Achillea millefolium
  Arctium lappa
  Brassica nigra
  Chrysanthemum leucanthemum
  Glechoma hederacea
  Matricaria recutita
  Urtica dioica

Sedatives
  Berberis vulgaris
  Galium aparine
  Matricaria recutita
  Verbascum thapsus

Shingles *see* HERPES

Skin Disorders
  Achillea millefolium
  Agropyron repens
  Arctium lappa
  Berberis vulgaris
  Capsella bursa-pastoris

  Chelidonium majus
  Chrysanthemum leucanthemum
  Cichorium intybus
  Daucus carota
  Equisetum arvense
  Galium aparine
  Glechoma hederacea
  Matricaria recutita
  Nasturtium officinale
  Phytolacca americana
  Plantago major
  Rhus glabra
  Rhus radicans
  Rumex acetosella
  Rumex crispus
  Stellaria media
  Taraxacum officinale
  Tussilago farfara
  Urtica dioica
  Verbascum thapsus

Skin Ulcers *see* SORES/SKIN ULCERS

Snakebites
  Arctium lappa
  Galium aparine
  Plantago major
  Taraxacum officinale

Sores/Skin Ulcers
  Achillea millefolium
  Arctium lappa
  Berberis vulgaris
  Chenopodium album
  Daucus carota
  Galium aparine
  Glechoma hederacea

Matricaria recutita
Melilotus officinalis
Phytolacca americana
Plantago major
Rumex acetosella
Rumex crispus
Stellaria media
Tussilago farfara
Verbascum thapsus

Splenic Disorders

Agropyron repens
Berberis vulgaris
Rumex crispus
Taraxacum officinale

Splinters (extraction aid)

Plantago major
Verbascum thapsus

Sprains

Arctium lappa

Stimulants, Gastric

Achillea millefolium
Berberis vulgaris
Brassica nigra
Capsella bursa-pastoris
Daucus carota
Rhus radicans

Stomach Disorders *see* GASTRIC DISORDERS

Sunburn

Arctium lappa
Galium aparine

Swellings

Achillea millefolium
Capsella bursa-pastoris
Chenopodium album
Chrysanthemum leucanthemum
Cichorium intybus
Matricaria recutita
Melilotus officinalis
Rumex crispus
Verbascum thapsus

Throat Infections

Berberis vulgaris
Brassica nigra
Daucus carota
Phytolacca americana
Rhus glabra
Rumex acetosella
Rumex crispus
Stellaria media
Tussilago farfara
Verbascum thapsus

Thrush

Rumex crispus
Senecio vulgaris

Tonics

Achillea millefolium
Agropyron repens
Arctium lappa
Berberis vulgaris
Brassica nigra
Capsella bursa-pastoris
Chrysanthemum leucanthemum
Cichorium intybus
Equisetum arvense

Galium aparine
Glechoma hederacea
Matricaria recutita
Nasturtium officinale
Phytolacca americana
Plantago major
Rhus glabra
Rumex crispus
Stellaria media
Taraxacum officinale
Tussilago farfara
Urtica dioica
Verbascum thapsus

Toothache

Achillea millefolium
Brassica nigra
Chelidonium majus
Matricaria recutita
Plantago major
Portulaca oleracea
Rumex crispus
Urtica dioica
Verbascum thapsus

Tuberculosis

Chrysanthemum leucanthemum
Glechoma hederacea
Taraxacum officinale
Urtica dioica

Tumors

Achillea millefolium
Arctium lappa
Chelidonium majus
Equisetum arvense
Galium aparine
Melilotus officinalis

Ulcers, Stomach

Daucus carota

Upper Respiratory Infections

Achillea millefolium
Agropyron repens
Arctium lappa
Brassica nigra
Chrysanthemum leucanthemum
Daucus carota
Equisetum arvense
Galium aparine
Glechoma hederacea
Matricaria recutita
Nasturtium officinale
Phytolacca americana
Plantago major
Portulaca oleracea
Rumex acetosella
Rumex crispus
Stellaria media
Taraxacum officinale
Tussilago farfara
Urtica dioica
Verbascum thapsus

Urinary Disorders

Achillea millefolium
Agropyron repens
Capsella bursa-pastoris
Cichorium intybus
Daucus carota
Equisetum arvense
Galium aparine
Glechoma hederacea
Matricaria recutita
Rhus glabra

Rumex acetosella
Senecio vulgaris
Taraxacum officinale
Urtica dioica

Urine Incontinence

Achillea millefolium
Capsella bursa-pastoris
Equisetum arvense
Rhus radicans
Taraxacum officinale
Urtica dioica
Verbascum thapsus

Varicose Ulcers

Daucus carota
Galium aparine
Plantago major

Varicose Veins

Achillea millefolium
Capsella bursa-pastoris

Venereal Disease

Rhus glabra

Warts

Chelidonium majus

Taraxacum officinale
Verbascum thapsus

Worms, Intestinal

Daucus carota
Melilotus officinalis
Plantago major
Portulaca oleracea
Senecio vulgaris

Wounds, Large/Open

Achillea millefolium
Arctium lappa
Berberis vulgaris
Capsella bursa-pastoris
Chelidonium majus
Chrysanthemum leucanthemum
Equisetum arvense
Galium aparine
Glechoma hederacea
Matricaria recutita
Nasturtium officinale
Plantago major
Rumex acetosella
Rumex crispus
Senecio vulgaris
Tussilago farfara
Urtica dioica
Verbascum thapsus

APPENDIX B: HERBAL AUTHORITIES THROUGH THE AGES

PHYSICIANS, BOTANISTS, NATURALISTS, HERBALISTS, APOTHECARIES

NAME	DATES	NATIONALITY
Hippocrates	460(?)–377(?) B.C.	Greek physician, "Father of Medicine"
Theophrastus	371/370–288/287 B.C.	Greek philosopher, naturalist
Dioscorides, Pedanios	First Century A.D.	Greek medical writer (*Materia medica*)
Pliny, the Elder (Gaius Plinius Secundus)	A.D. 23/24–79	Roman naturalist, author (*Historia naturalis*)
Galen (Galenos, Claudius)	A.D. 130–200	Greek physician, medical writer
St. Hildegard von Bingen	1098–1179	German mystic, naturalist
Ruellius, Joannes (Jean Ruel)	1474–1537	French physician
Brunfels, Otto	1489–1534	German physician, botanist, writer
Monardes, Nicolas	1493–1578	Spanish physician
Paracelsus (Philippus Aureolus Theophrastus Bombast von Hohenheim)	1493(?)–1541	Swiss chemist

NAME	DATES	NATIONALITY
Bock, Hieronymus	1498–1554	German Protestant priest
Fuchs, Leonhart	1501–1566	German physician
Matthiolus, Petrus Andreas (Pierandrea Mattioli)	1501–1577	Italian lawyer, physician
Turner, William	1510?–1568	"Father of English Botany"
Gesner, Konrad	1516–1565	Swiss naturalist
Dodonaeus (Dodoens, Rembert)	1517–1585	Flemish physician, botanist
Tabernaemontanus (Jakob Theodor von Bergzabern)	1520–1590	German physician
Clusius, Carolus (Charles de l'Ecluse)	1526–1609	Dutch botanist
Lobelius (Mathias de l'Obel)	1538–1616	Flemish physician, botanist
Gerard, John	1545–1612	English surgeon, apothecary
Parkinson, John	1567–1650	English botanist, herbalist, apothecary
Markham, Gervase	1568–1637	English naturalist, writer
Culpeper, Nicholas	1616–1654	English apothecary, herbalist
Evelyn, John	1620–1706	English diarist
Josselyn, John	1630–1675	English traveler, author (*New England Rarities Discovered*)
Linnaeus, Carolus (Carl von Linné)	1707–1778	Swedish biologist
Cyrillo (Cirillo), Domenico Maria Leone	1739–1799	Italian statesman

NAME	DATES	NATIONALITY
Beauvois, Ambroise Marie François Joseph Palisot de	1752–1820	French explorer
Weber, Georg Heinrich	1752–1828	German botanist
Fourcroy, Antoine François Count de	1755–1809	French chemist
Hahnemann, Samuel	1755–1843	German chemist-physician, founder of homeopathy
Nuttall, Thomas	1786–1859	English botanist
Bentham, George	1800–1884	English botanist, plant cataloguer
Kneipp, Father Sebastian	1821–1897	German herbalist, advocate of water cures
Künzle, Father Johann	1857–1945	Swiss "Kräuterpfarrer" (herbal priest)
Hutchinson, John	1884–?	English botanist
Leclerc, Henri	Nineteenth Century	French botanist, physician

GLOSSARY OF MEDICAL TERMS AND HERBAL PREPARATIONS

ACRID	Leaving a burning sensation in the mouth and throat.
ALTERATIVE	Alters and corrects toxic conditions of the bloodstream, restores healthy functioning.
AMENORRHEA	Absence of menstruation.
ANEMIA	Deficiency in the number of red blood corpuscles.
ANTHELMINTIC	Expels or destroys parasitic worms, especially in the intestines.
ANTIBACTERIAL	Effective against bacteria.
ANTICATARRHAL	Effective against catarrh.
ANTICONVULSANT	Effective against convulsions.
ANTIDOTE	An agent that counteracts or destroys poison.
ANTIPYRETIC	Reduces fever.
ANTISCORBUTIC	Counteracts scurvy, which results from a lack of vitamin C.
ANTISEPTIC	Destroys or inhibits bacteria, helps to counteract putrefaction.
ANTISPASMODIC	Counteracts or eases spasms or cramps.
APERIENT	Gently moving the bowels.
APHRODISIAC	Stimulates sexual desire.
APPETIZER	Stimulates the appetite.
ARTERIOSCLEROSIS	Hardening of arterial walls.
ARTHRITIS	Inflammation of joints.
ASTHMA	Wheezing, gasping, severe shortness of breath.
ASTRINGENT	Causes contraction of injured tissue.
BACTERICIDE	An agent that destroys bacteria.

BALSAMIC	Aromatic, resinous odor or remedy.
CARDIAC	Acting upon the heart.
CARMINATIVE	Expelling gas from alimentary canal.
CATHARTIC	Stronger than aperient, laxative. See also *Purgative*
CHOLAGOGUE	Promoting the flow of bile from the body's system.
COMPRESS	Folded cloth or pad, moistened with infusion, applied to a body part.
COUNTERIRRITANT	An agent applied locally to the skin to reduce inflammation in deeper areas. See also *Rubefacient*
DECOCTION	Extracting flavor and essential oils from an herb, by boiling. See also *Infusion*
DEMULCENT	Bland or oily substance to soothe iritated mucous membranes.
DEOBSTRUENT	Removing obstructions from ducts in the body.
DEPURATIVE	Tending to cleanse; a purifying substance.
DETERGENT	Cleansing agent.
DIAPHORETIC	Promoting increased perspiration.
DIGESTIVE	Aids digestion of food.
DIURETIC	Able to increase the flow of urine.
EMETIC	An agent that induces vomiting.
EMMENAGOGUE	Promotes menstrual discharge.
EMOLLIENT	Soothes skin or mucous membranes.
EXPECTORANT	Promoting discharge of phlegm from throat or lungs by coughing.
FEBRIFUGE	Reduces fever.
FLATULENCE	Gases generated in stomach or intestines.
GERMICIDE	An agent that destroys germs.
HEMOSTATIC	Checks bleeding.
HERBAL BATH	Bathwater to which an herbal decoction has been added.
INFUSION	Tea prepared by pouring boiling water (1 cup) on an herb (1 teaspoon fresh; ½ teaspoon dry) to release the herb's flavor and essential oils.
LAXATIVE	Relieves constipation.
LITHONTRIPTIC	An agent for dissolving or destroying stones in bladder or kidneys.

LOTION	A liquid preparation, e.g. infusion, decoction, with or without added ingredients.
NERVINE	Relieves nervous tension and excitement.
NEURALGIA	Acute nerve pain.
NUTRIENT, NUTRITIVE	Supplying nourishment.
OILS AND OINTMENTS	Four ounces fresh or 2 ounces dried bruised herbs in 1 pint vegetable oil, set in warm place 1 week, shaken daily, then strained and bottled; *or* same quantities bruised herbs in ½ pint vegetable oil and ½ pound rendered lard, *gently* heated, uncovered, one hour in enamel or glass pot; slightly cooled, strained, and potted.
PLASTER	A cloth spread with herbal paste, to be applied to a body part; paste sometimes "sandwiched" between two cloths.
POULTICE	Usually fresh crushed or heated fresh herbs applied alone, or first spread on cloth, and applied to sores, et al.
PURGATIVE	Produces copious evacuation from bowels.
REFRIGERANT	Reduces fever, body heat, thirst.
RESTORATIVE	Restores normal health, vigor.
RUBEFACIENT	Produces redness of the skin. See also *Counterirritant*
SEDATIVE	Induces calmness, relaxation.
STIMULANT	Increases the body system's functional activity.
STOMACHIC	Stimulates stomach functions.
STYPTIC	Tends to check bleeding. See also *Astringent*
SUDORIFIC	Causes or induces sweat.
TEA	See *Infusion*
TISANE	See *Infusion*; usually implies medicinal effects.
TONIC	An agent that invigorates, refreshes, stimulates, restores body.
VERMIFUGE	Expels or destroys parasitic worms.

BIBLIOGRAPHY

Andrews, Jonathan. Anthony Huxley, ed. *Creating a Wild Flower Garden.* New York: Henry Holt and Company, 1986.

Beedell, Suzanne. *Pick, Cook and Brew.* London: Pelham Books, 1973.

Berglund, Berndt, and Clare E. Bolsby. *The Edible Wild: A Complete Cookbook and Guide to Edible Wild Plants in Canada and Eastern North America.* New York: Charles Scribner's Sons, 1971.

Blunt, Wilfrid. *Art of Botanical Illustrations.* London: William Collins, 1950.

Branchini, Francesco, and Francesco Corbetta. *Health Plants of the World: Atlas of Medicinal Plants.* New York: Newsweek Books, 1977.

Buchman, Dian Dincin. *Feed Your Face: A Complete Herbal Guide to Natural Beauty and Health.* London: Gerald Duckworth and Company, 1973.

Clark, Linda. *Get Well Naturally.* Old Greenwich, Conn.. The Devin-Adair Company, 1967

Clarkson, Rosetta E. *Golden Age of Herbs and Herbalists.* New York: Dover Publications, 1972.

Clarkson, Rosetta E. *Herbs and Savory Seeds.* New York: Dover Publications, 1972.

Clarkson, Rosetta E. *Herbs: Their Culture and Uses.* New York: Macmillan Publishing Company, 1976.

Cobb, Boughton. *Ferns.* Peterson Field Guides. Boston: Houghton Mifflin Company, 1963.

Coon, Nelson. *Dictionary of Useful Plants.* Emmaus, Pa.: Rodale Press, 1974.

Coon, Nelson. *Using Plants for Healing.* Emmaus, Pa.: Rodale Press, 1979.

Creasy, Rosalind. *The Gardener's Handbook of Edible Plants.* San Francisco: Sierra Club Books, 1986.

Crowhurst, Adrienne. *The Weed Cookbook.* New York: Lancer Books, 1972.

Culpeper, Nicholas. *Culpeper's Complete Herbal: Consisting of a Comprehensive Description of Nearly All Herbs with Their Medicinal Properties and Directions for Compounding the Medicines Extracted from Them.* London: W. Foulsham and Company.

Densmore, Frances. *How Indians Use Wild Plants for Food, Medicine & Crafts.* New York: Dover Publications, 1974.

Ferguson, Nicola. *Right Plant, Right Place.* New York: Summit Books, 1984.

Fink-Henseler, Roland. *Grossmutters Hausapotheke.* Bindlach, Germany: Gondrom Verlag, 1986.

Fischer, Eugen. *Heilpflanzen.* Bern: Hallweg AG, 1982.

Freeman, Margaret B. *Herbs for the Mediaeval Household for Cooking, Healing and Divers Uses.* New York: The Metropolitan Museum of Art, 1943.

Gibbons, Euell. *Stalking the Healthful Herbs.* New York: David McKay Company, 1970.

Gibbons, Euell. *Stalking the Wild Asparagus.* New York: David McKay Company, 1962.

Glob, P. V. *The Bog People: Iron-Age Man Preserved.* New York: Ballantine Books, 1973.

Gray, Asa. *Manual of Botany.* 8th ed. Edited and revised by D. Fernold. New York: American Book Company, 1950.

Grieve, Mrs. M. *Culinary Herbs & Condiments.* New York: Dover Publications, 1971.

Grieve, Mrs. M. *A Modern Herbal: The Medicinal, Culinary, Cosmetic and Economic Properties, Cultivation and Folk-Lore of Herbs, Grasses, Fungi, Shrubs & Trees with All Their Modern Scientific Uses.* New York: Dover Publications, 1971.

Griggs, Barbara. *The Home Herbal: A Handbook of Simple Remedies.* London: Pan Books, 1986.

Harper-Shove, F. *The Prescriber and Clinical Repertory of Medicinal Herbs.* 2d ed. London: Homeopathic Publishing Company, 1938.

Harrop, Renny, ed. *The Book of Herbs.* New York: Exeter Books, 1985.

Hill, Albert. *Economic Botany.* New York: McGraw-Hill Book Company, 1937.

Hutchens, Alma R. *Indian Herbalogy of North America.* Windsor, Ontario, Canada: MERCO, 1973.

Hylton, William H., ed. *The Rodale Herb Book.* Emmaus, Pa.: Rodale Press, 1974.

Josselyn, John. *New-England Rarities Discovered.* London: N.p. 1672.

Kingsbury, John M. *Deadly Harvest: A Guide to Common Poisonous Plants.* New York: Holt, Rinehart and Winston, 1965.

Kingsbury, John M. *Poisonous Plants of the United States and Canada.* Englewood Cliffs, N.J.: Prentice-Hall, 1964.

Klimas, John E., and James A. Cunningham. *Wildflowers of Eastern America.* New York: Alfred A. Knopf, 1974.

Kneipp, Sebastian. *Meine Wasserkur, So sollt Ihr leben: Die weltberühmten Ratgeber in einem Band.* Munich: Ehrenwirth Verlag GmbH, 1988.

Kowalchik, Claire, and William H. Hylton, eds. *Rodale's Illustrated Encyclopedia of Herbs.* Emmaus, Pa.: Rodale Press, 1987.

Kraus, Barbara. *Calories and Carbohydrates.* New York: New American Library, 1975.

Krochmal, Arnold, and Annie Krochmal. *A Field Guide to Medicinal Plants.* New York: Time Books, 1984.

Künzle, Johann. *Das grosse Kräuterheilbuch.* Olten, Switzerland: Walter-Verlag, 1945.

Lewis, Walter H., et al. *Plants Affecting Man's Health.* New York: John Wiley and Sons, 1977.

Lucas, Richard. *Common & Uncommon Uses of Herbs for Healthful Living.* New York: Arco Publishing Company, 1969.

Magic and Medicine of Plants. Pleasantville, New York: The Reader's Digest Association, 1986.

Marks, Geoffrey, and William K. Beatty. *The Medical Garden.* New York: Charles Scribner's Sons, 1973.

Mathews, F. Schuyler. *Field Book of American Wild Flowers.* 1902. Rev. ed. New York: G. P. Putnam and Sons, 1955.

Medsger, Oliver Perry. *Edible Wild Plants.* New York: Macmillan Publishing Company, 1966.

Muenscher, Walter C. L. *Poisonous Plants of the United States.* New York: Macmillan Publishing Company, 1964.

Northcote, Lady Rosalind. *The Book of Herb Lore*. New York: Dover Publications, 1971 (orig. publ. 1912).

Palaiseul, Jean. *Grandmother's Secrets: Her Green Guide to Health from Plants*. Harmondsworth, Middlesex, England: Penguin Books, 1986.

Peterson, Lee Allen. *Edible Wild Plants*. Peterson Field Guides. Boston: Houghton Mifflin Company, 1977.

Peterson, Roger Tory, and Margaret McKenny. *A Field Guide to Wildflowers*. Boston: Houghton Mifflin Company, 1968.

Pfeiffer, Ehrenfried E. *Weeds and What They Tell*. Wyoming, R.I.: Bio-Dynamic Literature. Reprinted by permission of Rodale Press, Emmaus, Pa.: 1981.

Prentice, Thurlow Merrill. *Weeds and Wildflowers of Eastern North America*. Salem: Peabody Museum, 1973.

Quelch, Mary Thorne. *Herbs for Daily Use*. London: Faber and Faber, 1957.

Saunders, Charles Francis. *Useful Wild Plants of the United States and Canada*. New York: Robert M. McBride Company, 1934.

Schwester Bernardines Hausmittelbuch. Munich: Mosaik Verlag GmbH, 1982.

Smith, William. *Wonders in Weeds*. Saffron Walden, England: The C. W. Daniel Company, 1983.

Spencer, Edwin Rollin. *All About Weeds*. New York: Dover Publications, 1974.

Stein, Sara B. *My Weeds*. New York: Harper & Row Publishers, 1988.

Stuart, Malcolm, ed. *The Encyclopedia of Herbs and Herbalism*. New York: Crescent Books, 1987.

Szczawinski, Adam F. *Edible Garden Weeds of Canada*. Ottawa: National Museum of Natural Sciences, 1978.

Thomas, Mai. *Grannies' Remedies*. New York: James H. Heineman, 1967.

Thomson, William A. R., M.D. *Herbs That Heal*. New York: Charles Scribner's Sons, 1976.

Thomson, William A. R., M.D. *Medicines from the Earth: A Guide to Healing Plants*. San Francisco: Harper & Row, 1983.

Tierra, Michael. *The Way of Herbs*. New York: Pocket Books, 1983.

Tobe, John H. *Proven Herbal Remedies.* New York: Pyramid Publications, 1973.

Treben, Maria. *Heilerfolge zum Buch Heilkräuter aus dem Garten Gottes.* Munich: Wilhelm Heyne Verlag, 1987.

Tyler, Varro E., Ph.D. *The Honest Herbal.* Philadelphia: George F. Stickley Company, 1982.

Vogel, Alfred. *Swiss Nature Doctor.* Teufen, Switzerland: Edition A. Vogel, 1980.

Weiner, Michael. *Earth Medicine—Earth Foods: Plant Remedies, Drugs, and Natural Foods of the North American Indians.* First revised and expanded edition. New York: Macmillan Publishing Company, 1980.

Weiner, Michael A., M.S., M.A., Ph.D., with Janet Weiner. *Weiner's Herbal.* Briarcliff Manor, N.Y.: Stein and Day, 1980.

Woodward, Marcus. *Leaves from Gerard's Herball.* New York: Dover Publications, 1969.

Wren, R. C. *Potter's New Cyclopaedia of Botanical Drugs and Preparations.* 3d ed. London: Potter and Clarke, 1923.

INDEX

OF

BOTANICAL AND

COMMON NAMES

Aaron's Rod <u>see</u> *Verbascum thapsus*

Adam's Flannel <u>see</u> *Verbascum thapsus*

Agleaf <u>see</u> *Verbascum thapsus*

Alehoof <u>see</u> *Glechoma hederacea*

Allgood <u>see</u> *Chenopodium album*

Alsine media <u>see</u> *Stellaria media*

American Cancer <u>see</u> *Phytolacca americana*

American Nightshade <u>see</u> *Phytolacca americana*

American Spinach <u>see</u> *Phytolacca americana*

Angel Flower <u>see</u> *Achillea millefolium*

Anthemis nobilis <u>see</u> *Matricaria recutita*

Baccy Plant <u>see</u> *Tussilago farfara*

Baconweed <u>see</u> *Chenopodium album*

Bad Man's Plaything <u>see</u> *Achillea millefolium*

Barbe-de-Capucin <u>see</u> *Cichorium intybus*

Barberry <u>see</u> *Berberis vulgaris*

Bardane <u>see</u> *Arctium lappa*

Barweed <u>see</u> *Galium aparine*

Bear's Grape <u>see</u> *Phytolacca americana*

Bedstraw <u>see</u> *Galium aparine*

Bee's Nest <u>see</u> *Daucus carota*

Beggar's Blanket <u>see</u> *Verbascum thapsus*

Beggar's Buttons <u>see</u> *Arctium lappa*

Berbery <u>see</u> *Berberis vulgaris*

Bird Seed <u>see</u> *Plantago major*

Bird Seed <u>see</u> *Senecio vulgaris*

Bird's Nest <u>see</u> *Daucus carota*

Bitter Dock <u>see</u> *Rumex crispus*

Black Mustard <u>see</u> *Brassica nigra*

Blanket Herb <u>see</u> *Verbascum thapsus*

Bloodwort <u>see</u> *Achillea millefolium*

Blowball <u>see</u> *Taraxacum officinale*

Blue Dandelion <u>see</u> *Cichorium intybus*

Blue Runner <u>see</u> *Glechoma hederacea*

Blue Sailors <u>see</u> *Cichorium intybus*

Bokhara Clover <u>see</u> *Melilotus officinalis*

Bottlebrush <u>see</u> *Equisetum arvense*

Brassica hirta <u>see</u> *Brassica nigra*

Brassica juncea <u>see</u> *Brassica nigra*

Broad-Leaved Plantain <u>see</u> *Plantago major*

Brown Mustard <u>see</u> *Brassica nigra*

Bull Daisy <u>see</u> *Chrysanthemum leucanthemum*

Bull Pipes <u>see</u> *Equisetum arvense*

Bullsfoot <u>see</u> *Tussilago farfara*

Bunch o' Daisies <u>see</u> *Achillea millefolium*

Dung Weed <u>see</u> *Chenopodium album*

Dutch Grass <u>see</u> *Agropyron repens*

Dutch Morgan <u>see</u> *Chrysanthemum leucanthemum*

Dutch Rushes <u>see</u> *Equisetum arvense*

Earth Ivy <u>see</u> *Glechoma hederacea*

Englishman's Foot <u>see</u> *Plantago major*

Eriffe <u>see</u> *Galium aparine*

European Barberry <u>see</u> *Berberis vulgaris*

Everlasting Friendship <u>see</u> *Galium aparine*

Fair-Maids-of-France <u>see</u> *Chrysanthemum leucanthemum*

Fat Hen <u>see</u> *Chenopodium album*

Felonwort <u>see</u> *Chelidonium majus*

Feltwort <u>see</u> *Verbascum thapsus*

Field Balm <u>see</u> *Glechoma hederacea*

Field Daisy <u>see</u> *Chrysanthemum leucanthemum*

Field Horsetail <u>see</u> *Equisetum arvense*

Fieldhove <u>see</u> *Tussilago farfara*

Field Sorrel <u>see</u> *Rumex acetosella*

Filius ante patrem <u>see</u> *Tussilago farfara*

Fireweed <u>see</u> *Plantago major*

Flannel Leaf <u>see</u> *Verbascum thapsus*

Flos cuculi <u>see</u> *Nasturtium officinale*

Fluffweed <u>see</u> *Verbascum thapsus*

Foalswort <u>see</u> *Tussilago farfara*

Fox's Clote <u>see</u> *Arctium lappa*

French Sorrel <u>see</u> *Rumex acetosella*

Frost Blite <u>see</u> *Chenopodium album*

Garden Celandine <u>see</u> *Chelidonium majus*

Garden Sorrel <u>see</u> *Rumex acetosella*

Garget <u>see</u> *Phytolacca americana*

German Camomile <u>see</u> *Matricaria recutita*

Gill-Creep-by-Ground <u>see</u> *Glechoma hederacea*

Gill-Go-by-the-Hedge <u>see</u> *Glechoma hederacea*

Gill-Go-over-the-Ground <u>see</u> *Glechoma hederacea*

Gobo <u>see</u> *Arctium lappa*

Golden Rod <u>see</u> *Verbascum thapsus*

Goldens <u>see</u> *Chrysanthemum leucanthemum*

Goosebill <u>see</u> *Galium aparine*

Goosefoot <u>see</u> *Chenopodium album*

Goose Grass <u>see</u> *Galium aparine*

Gooseheriff <u>see</u> *Galium aparine*

Goose Tongue <u>see</u> *Achillea millefolium*

Gowan <u>see</u> *Chrysanthemum leucanthemum*

Great Burdock <u>see</u> *Arctium lappa*

Great Mullein <u>see</u> *Verbascum thapsus*

Great Oxeye <u>see</u> *Chrysanthemum leucanthemum*

Greater Celandine <u>see</u> *Chelidonium majus*

Greater Plantain <u>see</u> *Plantago major*

Grecian May <u>see</u> *Chelidonium majus*

Greensauce <u>see</u> *Rumex acetosella*

Gripgrass <u>see</u> *Galium aparine*

Ground Glutton <u>see</u> *Senecio vulgaris*

Ground Ivy <u>see</u> *Glechoma hederacea*

Groundsel <u>see</u> *Senecio vulgaris*

Grundy Swallow <u>see</u> *Senecio vulgaris*

Gypsy's Rhubarb <u>see</u> *Arctium lappa*

Hag's Taper <u>see</u> *Verbascum thapsus*

Happy-Major <u>see</u> *Arctium lappa*

Hardock <u>see</u> *Arctium lappa*

Harebur <u>see</u> *Arctium lappa*

Hare-Lock <u>see</u> *Arctium lappa*

Hare's Beard <u>see</u> *Verbascum thapsus*

Hart's Clover <u>see</u> *Melilotus officinalis*

Hart's Tree <u>see</u> *Melilotus officinalis*

Haymaids <u>see</u> *Glechoma hederacea*

Hayriffe <u>see</u> *Galium aparine*

Healing Blade <u>see</u> *Plantago major*

Heart-Fever-Grass <u>see</u> *Taraxacum officinale*

Three-Leaved Ivy <u>see</u> *Rhus radicans*
Tongue Grass <u>see</u> *Stellaria media*
Torches <u>see</u> *Verbascum thapsus*
Torchwort <u>see</u> *Verbascum thapsus*
Touch-Me-Not <u>see</u> *Rhus radicans*
Toywort <u>see</u> *Capsella bursa-pastoris*
Triticum repens <u>see</u> *Agropyron repens*
Tunhoof <u>see</u> *Glechoma hederacea*
Turkey-Bur <u>see</u> *Arctium lappa*
Twitch Grass <u>see</u> *Agropyron repens*
Upland Sumac <u>see</u> *Rhus glabra*
Velvet Dock <u>see</u> *Verbascum thapsus*
Virginia Poke <u>see</u> *Phytolacca americana*
Wartweed <u>see</u> *Chelidonium majus*
Watercress <u>see</u> *Nasturtium officinale*
Waybread <u>see</u> *Plantago major*
White Golds <u>see</u> *Chrysanthemum leucanthemum*
White Man's Foot <u>see</u> *Plantago major*
White Mustard <u>see</u> *Brassica nigra*

Whiteweed <u>see</u> *Chrysanthemum leucanthemum*
Wild Camomile <u>see</u> *Matricaria recutita*
Wild Carrot <u>see</u> *Daucus carota*
Wild Endive <u>see</u> *Cichorium intybus*
Wild Ice Leaf <u>see</u> *Verbascum thapsus*
Wild Laburnum <u>see</u> *Melilotus officinalis*
Wild Pepper <u>see</u> *Achillea millefolium*
Wild Rhubarb <u>see</u> *Tussilago farfara*
Wild Sorrel <u>see</u> *Rumex acetosella*
Wild Spinach <u>see</u> *Chenopodium album*
Winter Weed <u>see</u> *Stellaria media*
Witches' Pouches <u>see</u> *Capsella bursa-pastoris*
Witch Grass <u>see</u> *Agropyron repens*
Witloof <u>see</u> *Cichorium intybus*
Yarrow <u>see</u> *Achillea millefolium*
Yarroway <u>see</u> *Achillea millefolium*
Yellow Dock <u>see</u> *Rumex crispus*
Yellow Gowan <u>see</u> *Taraxacum officinale*
Yellow Sweet Clover <u>see</u> *Melilotus officinalis*

INDEX

ABOUT THE AUTHOR

Pamela Jones, educated in Germany and England, moved to New York City where she had a career in publishing and public relations for many years. Giving up urban anxiety in 1980, she became a professional gardener and landscape designer in the rural tranquillity of upstate New York. Ms. Jones is also the author of *Under the City Streets: The History of Subterranean New York* and *How Does Your Garden Grow?: The Essential Home Garden Book.*